IMPACTS OF A WARMING ARCTIC
ARCTIC CLIMATE IMPACT
ACIA
ASSESSMENT
CAMBRIDGE
UNIVERSITY PRESS
D0723016

PUBLISHED BY THE PRESS SYNDICATE OF THE UNIVERSITY OF CAMBRIDGE
The Pitt Building, Trumpington Street, Cambridge, United Kingdom

CAMBRIDGE UNIVERSITY PRESS
The Edinburgh Building, Cambridge, CB2 2RU, UK
40 West 20th Street, New York, NY 10011-4211, USA
10 Stamford Road, Oakleigh, VIC 3166, Australia
Ruiz de Alarcón 13, 28014 Madrid, Spain
Dock House, The Waterfront, Cape Town 8001, South Africa

http://www.cambridge.org

First published 2004

Printed in Canada

ISBN 0 521 61778 2 paperback

AMAP Secretariat
P.O. Box 8100 Dep.
N-0032 Oslo, Norway
Tel: +47 23 24 16 30
Fax: +47 22 67 67 06
http://www.amap.no

CAFF International Secretariat
Hafnarstraeti 97
600 Akureyri, Iceland
Tel: +354 461-3352
Fax: +354 462-3390
http://www.caff.is

IASC Secretariat
Middelthuns gate 29
P.O. Box 5156 Majorstua
N-0302 Oslo, Norway
Tel: +47 2295 9900
Fax: +47 2295 9901
http://www.iasc.no

Author
Susan Joy Hassol

Project Production and Graphic Design
Paul Grabhorn, Joshua Weybright, Clifford Grabhorn (Cartography)

Photography
Bryan and Cherry Alexander, and others: credits on page 139

Technical editing
Carolyn Symon

Contributors

Assessment Integration Team

Robert Corell, Chair	American Meteorological Society, USA
Pål Prestrud, Vice Chair	Centre for Climate Research in Oslo, Norway
Gunter Weller	University of Alaska Fairbanks, USA
Patricia A. Anderson	University of Alaska Fairbanks, USA
Snorri Baldursson	Liaison for the Arctic Council, Iceland
Elizabeth Bush	Environment Canada, Canada
Terry V. Callaghan	Abisko Scientific Research Station, Sweden Sheffield Centre for Arctic Ecology, UK
Paul Grabhorn	Grabhorn Studio, Inc., USA
Susan Joy Hassol	Independent Scholar and Science Writer, USA
Gordon McBean	University of Western Ontario, Canada
Michael MacCracken	Climate Institute, USA
Lars-Otto Reiersen	Arctic Monitoring and Assessment Programme, Norway
Jan Idar Solbakken	Permanent Participants, Norway

ACIA Secretariat
Gunter Weller, Executive Director
Patricia A. Anderson, Deputy Executive Director
Barb Hameister, Sherry Lynch
International Arctic Research Center
University of Alaska Fairbanks
Fairbanks, AK 99775-7740, USA
Tel: +907 474 5818
Fax +907 474 6722
http://www.acia.uaf.edu

Lead Authors of the Full Science Report

Jim Berner	Alaska Native Tribal Health Consortium, USA
Terry V. Callaghan	Abisko Scientific Research Station, Sweden Sheffield Centre for Arctic Ecology, UK
Shari Fox	University of Colorado at Boulder, USA
Christopher Furgal	Laval University, Canada
Alf Håkon Hoel	University of Tromsø, Norway
Henry Huntington	Huntington Consulting, USA
Arne Instanes	Instanes Consulting Engineers, Norway
Glenn P. Juday	University of Alaska Fairbanks, USA
Erland Källén	Stockholm University, Sweden
Vladimir M. Kattsov	Voeikov Main Geophysical Observatory, Russia
David R. Klein	University of Alaska Fairbanks, USA
Harald Loeng	Institute of Marine Research, Norway
Marybeth Long Martello	Harvard University, USA
Gordon McBean	University of Western Ontario, Canada
James J. McCarthy	Harvard University, USA
Mark Nuttall	University of Aberdeen, Scotland, UK University of Alberta, Canada
Terry D. Prowse	National Water Research Institute, Canada
James D. Reist	Fisheries and Oceans Canada, Canada
Amy Stevermer	University of Colorado at Boulder, USA
Aapo Tanskanen	Finnish Meteorological Institute, Finland
Michael B. Usher	University of Stirling, Scotland, UK
Hjálmar Vilhjálmsson	Marine Research Institute, Iceland
John E. Walsh	University of Alaska Fairbanks, USA
Betsy Weatherhead	University of Colorado at Boulder, USA
Gunter Weller	University of Alaska Fairbanks, USA
Fred J. Wrona	National Water Research Institute, Canada

Note: A full list of additional contributors is found on page 129.

This report was prepared in English and translated into several other languages; the English version constitutes the official version.

Recommended Citation: ACIA, *Impacts of a Warming Arctic: Arctic Climate Impact Assessment.* Cambridge University Press, 2004.

http://www.acia.uaf.edu

Preface

The Arctic is of special importance to the world and it is changing rapidly. It is thus essential that decision makers have the latest and best information available regarding ongoing changes in the Arctic. This report is a plain language synthesis of the key findings of the Arctic Climate Impact Assessment (ACIA), designed to make the scientific findings accessible to policymakers and the broader public. The ACIA is a comprehensively researched, fully referenced, and independently reviewed evaluation of arctic climate change and its impacts for the region and for the world. It has involved an international effort by hundreds of scientists over four years, and also includes the special knowledge of indigenous people.

The Arctic Council called for this assessment, and charged two of its working groups, the Arctic Monitoring and Assessment Programme (AMAP) and the Conservation of Arctic Flora and Fauna (CAFF), along with the International Arctic Science Committee (IASC), with its implementation. Recognizing the central importance of the Arctic and this information to society as it contemplates responses to the growing challenge of climate change, the cooperating organizations are honored to forward this report to the Arctic Council and the international science community.

ACIA IS A PROJECT IMPLEMENTED BY AMAP, CAFF, AND IASC

The Arctic Council

The Arctic Council is a high-level intergovernmental forum that provides a mechanism to address the common concerns and challenges faced by arctic people and governments. It is comprised of the eight arctic nations (Canada, Denmark/Greenland/Faroe Islands, Finland, Iceland, Norway, Russia, Sweden, and the United States of America), six Indigenous Peoples organizations (Permanent Participants: Aleut International Association, Arctic Athabaskan Council, Gwich'in Council International, Inuit Circumpolar Conference, Russian Association of Indigenous Peoples of the North, and Saami Council), and official observers (including France, Germany, the Netherlands, Poland, United Kingdom, non-governmental organizations, and scientific and other international bodies).

The International Arctic Science Committee

The International Arctic Science Committee is a non-governmental organization whose aim is to encourage and facilitate cooperation in all aspects of arctic research among scientists and institutions of countries with active arctic research programs. IASC's members are national scientific organizations, generally academies of science, which seek to identify priority research needs, and provide a venue for project development and implementation.

Assessment Steering Committee

The ACIA Assessment Steering Committee was responsible for scientific oversight and coordination of all work related to the preparation of the assessment reports. A list of the members of this committee is found on page 138. The scientific content of the ACIA has been published in two separate reports: this synthesis volume, and a more comprehensive and detailed technical volume that includes references to the scientific literature. AMAP, CAFF, and IASC have received written certification by the ACIA leadership and all lead authors that the final scientific report reflects their expert views, and that this synthesis report is fully consistent with the scientific volume.

How To Read This Report

In presenting these findings, the stated likelihood of particular impacts occurring is based on expert evaluation of results from multiple lines of evidence including field and laboratory experiments, observed trends, theoretical analyses, and model simulations.

Judgments of likelihood based on these inputs are indicated using a five-tier lexicon consistent with everyday usage (very unlikely, unlikely, possible, likely, and very likely). Confidence in results is highest at both ends of this scale. A conclusion that an impact "will" result is reserved for situations where experience and multiple methods of analysis all make clear that the consequence would follow inevitably from the projected change in climate. Although many details of how climate, environment, and society will evolve in the future remain uncertain, experts do have more confidence in some findings than others. The use of the lexicon is thus designed to convey the current state of scientific understanding.

The projected impacts described in this report are based on observed data and a moderate scenario of future warming, *not a worst-case scenario.* Compared to the full range of scenarios analyzed by the Intergovernmental Panel on Climate Change (IPCC), the primary scenario used in the ACIA analyses falls below the middle of the IPCC range of projected temperature rise.

The results summarized in this report, like the extensive, fully referenced technical report upon which it is based, do not include thorough economic analyses of climate change impacts because the necessary information is not presently available. While adaptation strategies are sometimes mentioned, they are not analyzed in detail. The scope of this assessment did not include analysis of efforts to mitigate climate change impacts by reducing emissions of greenhouse gases.

A notation of which chapters of the full technical report have been principally drawn upon for the synthesis presented in this document is indicated in the bottom corner of each left-hand page (with the exception of the Executive Summary and Selected Sub-Regional Impacts, for which all chapters were drawn upon).

Finally, this assessment focused on impacts that are expected to occur within this century. Important longer-term impacts are occasionally mentioned, but not analyzed in detail.

Contents

Ice cores and other evidence of climate conditions in the distant past provide evidence that rising atmospheric carbon dioxide levels are associated with rising global temperatures. Human activities, primarily the burning of fossil fuels (coal, oil, and natural gas), and secondarily the clearing of land, have increased the concentration of carbon dioxide, methane, and other heat-trapping ("greenhouse") gases in the atmosphere. Since the start of the industrial revolution, the atmospheric carbon dioxide concentration has increased by about 35% and the global average temperature has risen by about 0.6°C. There is an international scientific consensus that most of the warming observed over the last 50 years is attributable to human activities.

"There is new and stronger evidence that most of the warming observed over the last 50 years is attributable to human activities."

Intergovernmental Panel on Climate Change (IPCC), 2001

Continuing to add carbon dioxide and other greenhouse gases to the atmosphere is projected to lead to significant and persistent changes in climate, including an increase in average global temperature of 1.4 to 5.8°C (according to the IPCC) over the course of this century. Climatic changes are projected to include shifts in atmospheric and oceanic circulation patterns, an accelerating rate of sea-level rise, and wider variations in precipitation. Together, these changes are projected to lead to wide-ranging consequences including significant impacts on coastal communities, animal and plant species, water resources, and human health and well-being.

About 80% of the world's energy is currently derived from burning fossil fuels, and carbon dioxide emissions from these sources are growing rapidly. Because excess carbon dioxide persists in the atmosphere for centuries, it will take at least a few decades for concentrations to peak and then begin to decline even if concerted efforts to reduce emissions are begun immediately. Altering the warming trend will thus be a long-term process, and the world will face some degree of climate change and its impacts for centuries.

The Earth's Greenhouse Effect

Most of the heat energy emitted from the surface is absorbed by greenhouse gases which radiate heat back down to warm the lower atmosphere and the surface. Increasing the concentrations of greenhouse gases increases the warming of the surface and slows the loss of heat energy to space.

The science suggests that responding to this challenge will require two sets of actions: one, called mitigation, to slow the speed and amount of future climate change by reducing greenhouse gas emissions; and the other, called adaptation, to attempt to limit adverse impacts by becoming more resilient to the climate changes that will occur while society pursues the first set of actions. The scope of this assessment did not include an evaluation of either of these sets of actions. These are being addressed by efforts under the auspices of the United Nations Framework Convention on Climate Change and other bodies.

Altering the warming trend will be a long-term process, and the world will face some degree of climate change and its impacts for centuries.

Stratospheric Ozone Depletion is Another Issue

Depletion of the stratospheric ozone layer due to chlorofluorocarbons and other man-made chemicals is a different problem, although there are important connections between ozone depletion and climate change. For example, climate change is projected to delay recovery of stratospheric ozone over the Arctic. This assessment, in addition to its principal focus on climate change impacts, also examined changes in stratospheric ozone, subsequent changes in ultraviolet radiation, and related impacts in the Arctic. A summary of these findings can be found on pages 98-105 of this report.

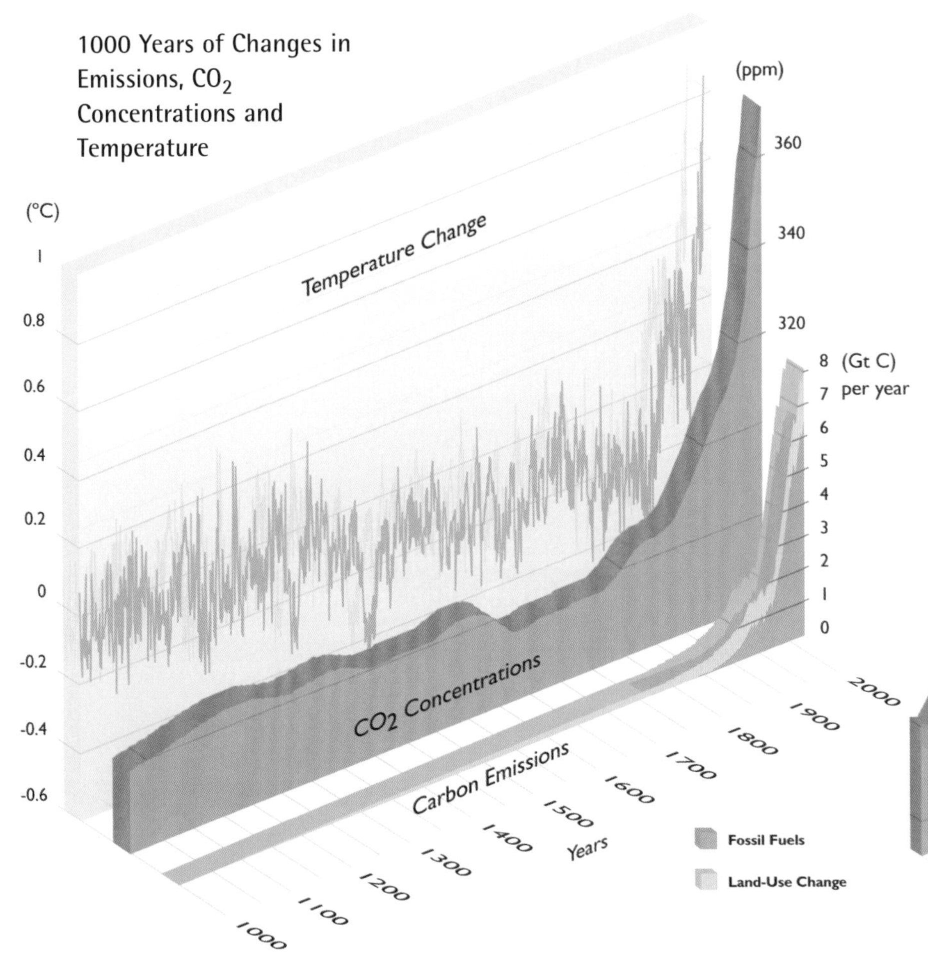

This 1000-year record tracks the rise in carbon emissions due to human activities (fossil fuel burning and land clearing) and the subsequent increase in atmospheric carbon dioxide concentrations and air temperatures. The earlier parts of this Northern Hemisphere temperature reconstruction are derived from historical data, tree rings, and corals, while the later parts were directly measured. Measurements of carbon dioxide (CO_2) in air bubbles trapped in ice cores form the earlier part of the CO_2 record; direct atmospheric measurements of CO_2 concentration began in 1957.

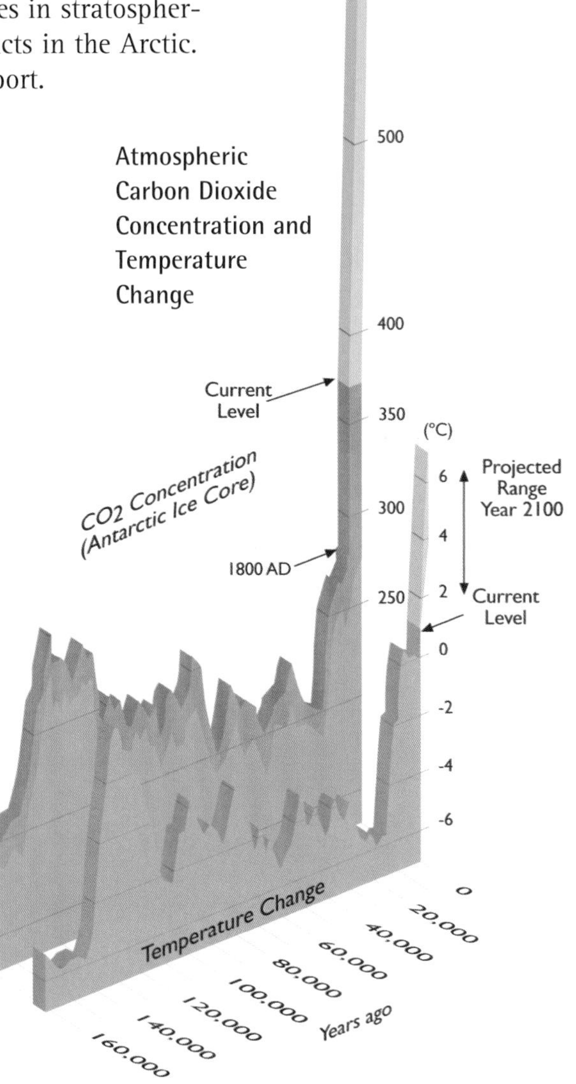

This record illustrates the relationship between temperature and atmospheric carbon dioxide concentrations over the past 160,000 years and the next 100 years. Historical data are derived from ice cores, recent data were directly measured, and model projections are used for the next 100 years.

The Arctic Region

Polaris, the North Star, is located almost directly above the North Pole. Around it are the stars that form the constellation known as Ursa Major, the Great Bear. The term Arctic comes from the ancient Greek word *Arktikós*, the country of the Great Bear.

Earth's northern polar region consists of a vast ocean surrounded by land, in contrast to the southern polar region in which an ice-covered continent is surrounded by ocean. Perhaps the most striking features are the snow and ice that cover much of the arctic land and sea surface, particularly in the high Arctic. And draped like a pair of great green shawls over the shoulders of the two facing continents are the boreal (meaning northern) forests. A wide expanse of tundra – treeless plains over frozen ground – lies between the icy high north and the forested sub-arctic.

One line often used to define the region is the Arctic Circle, drawn at the latitude north of which the sun does not rise above the horizon at winter solstice and does not set below it at summer solstice – "the land of the midnight sun". Other boundaries used to define the Arctic include treeline, climatic boundaries, and permafrost extent on land and sea-ice extent on the ocean. For the purposes of this assessment, the boundary will be more flexible, also encompassing sub-arctic areas integral to the functioning of the arctic system.

High arctic lands and seas are home to an array of plants, animals, and people that survive in some of the most extreme conditions on the planet. From the algae that live on the underside of sea ice, to the polar bears that hunt on top of the ice, to the indigenous human societies that have developed in close connection with their environment, these communities are uniquely adapted to what many outside the region would view as a very severe climate.

Life in the Arctic has historically been both vulnerable and resilient. Factors that contribute to the Arctic's vulnerability include its relatively short growing season and smaller variety of living things compared to temperate regions. In addition, arctic climate is highly variable, and a sudden summer storm or freeze can wipe out an entire generation of young birds, thousands of seal pups, or hundreds of caribou calves. Yet some arctic species have also displayed remarkable resilience to historic extremes, as evidenced by the recovery of populations that have occasionally been decimated by climatic variations.

The increasingly rapid rate of recent climate change poses new challenges to the resilience of arctic life. In addition to the impacts of climate change, many other stresses brought about by human activities are simultaneously affecting life in the Arctic, including air and water contamination, overfishing, increasing levels of ultraviolet radiation due to ozone depletion, habitat alteration and pollution due to resource extraction, and increasing pressure on land and resources related to the growing human population in the region. The sum of these factors threatens to overwhelm the adaptive capacity of some arctic populations and ecosystems.

The increasingly rapid rate of recent climate change poses new challenges to the resilience of arctic life.

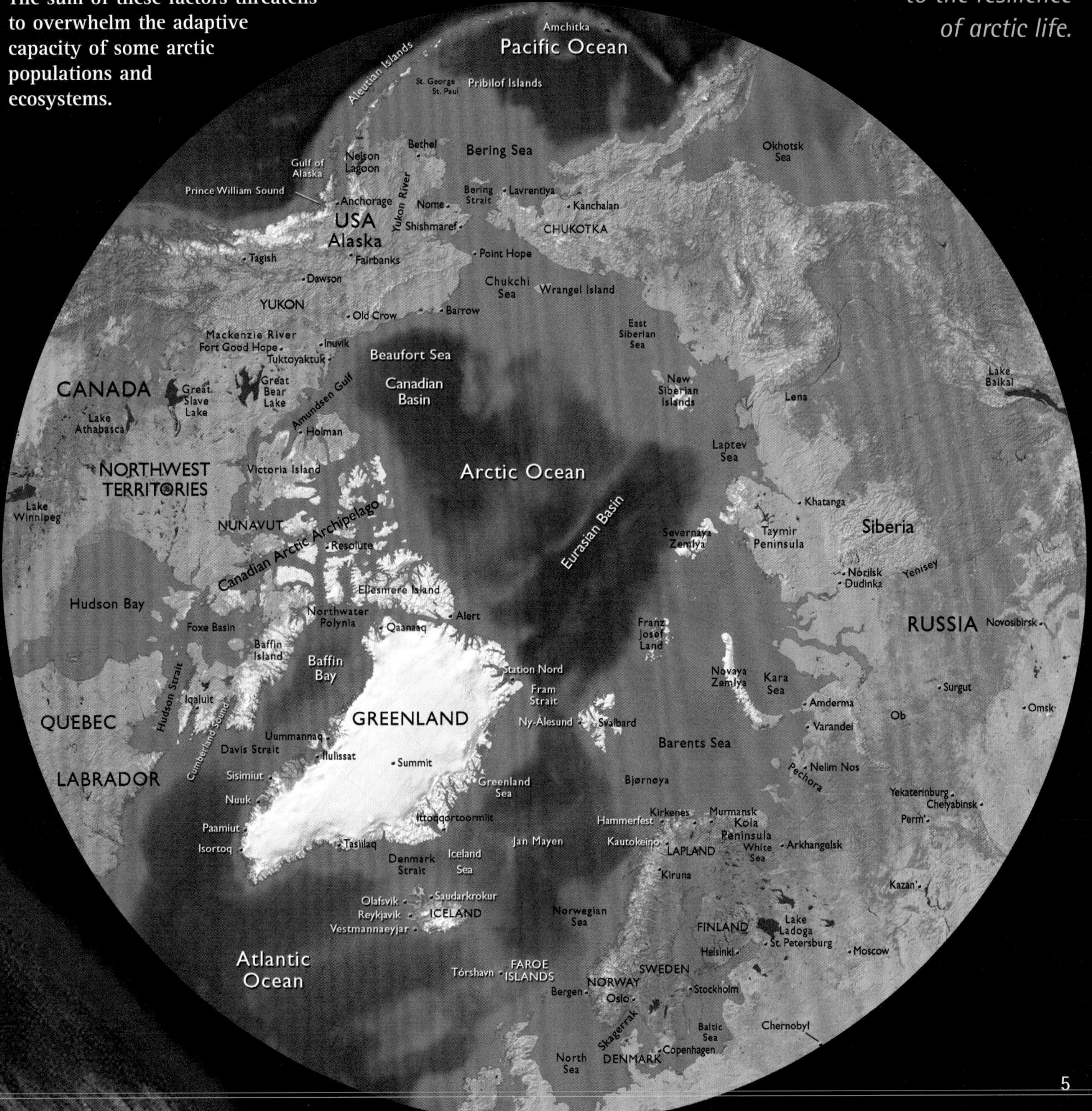

People of the Arctic

Almost four million people live in the Arctic today, with the precise number depending on where the boundary is drawn. They include indigenous people and recent arrivals, hunters and herders living on the land, and city dwellers. Many distinct indigenous groups are found only in the Arctic, where they continue traditional activities and adapt to the modern world at the same time. Humans have long been part of the arctic system, shaping and being shaped by the local and regional environment. In the past few centuries, the influx of new arrivals has increased pressure on the arctic environment through rising fish and wildlife harvests and industrial development.

The Arctic includes part or all of the territories of eight nations: Norway, Sweden, Finland, Denmark, Iceland, Canada, Russia, and the United States, as well as the homelands of dozens of indigenous groups that encompass distinct sub-groups and communities. Indigenous people currently make up roughly 10% of the total arctic population, though in Canada, they represent about half the nation's arctic population, and in Greenland they are the majority. Non-indigenous residents also include many different peoples with distinct identities and ways of life.

People have occupied parts of the Arctic since at least the peak of the last ice age, about 20,000 years ago, and recent studies suggest a human presence up to 30,000 years ago. In North America, humans are believed to have spread across the Arctic in several waves, reaching Greenland as many as 4500 years ago before abandoning the island for a millennium or more. Innovations such as the harpoon enabled people to hunt large marine mammals, making it possible to inhabit remote coastal areas in which the land offered scant resources. The development of reindeer husbandry in Eurasia allowed human populations to increase dramatically owing to a reliable food source. In Eurasia and across the North Atlantic, new groups of people moved northward over the past thousand years, colonizing new lands such as the Faroe Islands and Iceland, and encountering indigenous populations already present in West Greenland, and northern Norway, Sweden, Finland, and Russia.

In the 20th century, immigration to the Arctic increased dramatically, to the point where the non-indigenous population currently outnumbers the indigenous population in most regions. Many immigrants have been drawn by the prospect of new opportunities such as developing natural resources. Conflicts over land and resource ownership and access have been exacerbated by the rise in population and the incompatibility of some aspects of traditional and modern ways of life. In North America, the indigenous struggle to regain rights to land and resources has been addressed to some extent in land claims agreements, the creation of largely self-governed regions within nation-states, and other political and economic actions. In some areas, conflicts remain, particularly concerning the right to use living and mineral resources. In Eurasia, by contrast, indigenous claims and rights have only begun to be addressed as matters of national policy in recent years.

Populations are changing and northern regions are becoming more tightly related economically, politically, and socially to national mainstreams. Life expectancy has increased greatly across most of the Arctic in recent decades. The prevalence of indigenous language use, however, has decreased in most areas, with several languages in danger of disappearing in coming decades. In some respects, the disparities between northern and southern arctic communities in terms of living standards, income, and education are decreasing, although the gaps remain large in most cases.

Total and Indigenous Populations of the Arctic

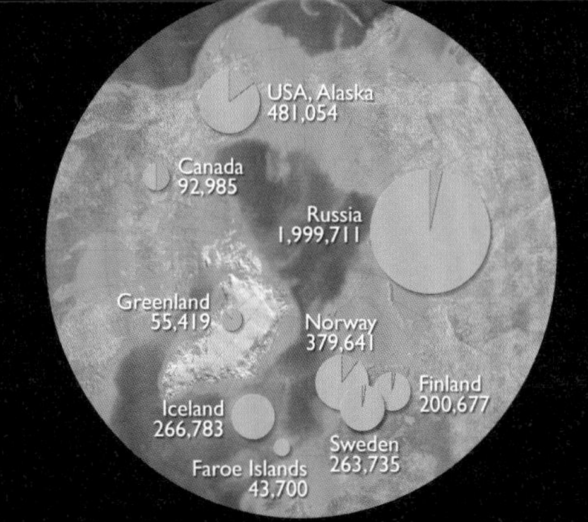

The economy of the region is based largely on natural resources, from oil, gas, and metal ores to fish, reindeer, caribou, whales, seals, and birds. In recent decades, tourism has added a growing sector to the economies of many communities and regions of the Arctic. Government services including the military are also a major part of the economy in nearly all areas of the Arctic, responsible in some cases for over half of the available jobs. In addition to the cash economy, traditional subsistence and barter economies make a major contribution to the overall well-being of parts of the region, producing significant value that is not recorded in official statistics.

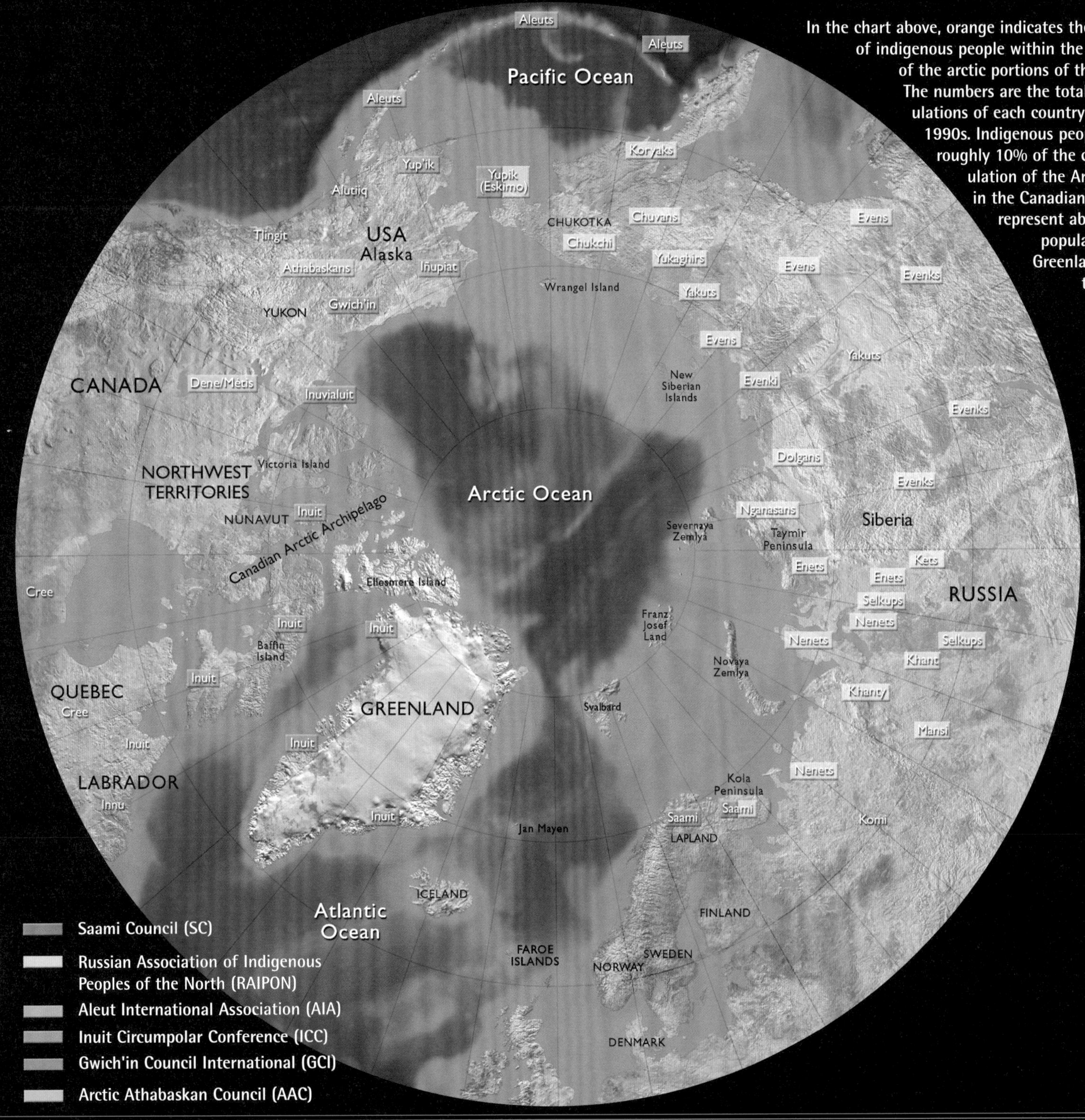

In the chart above, orange indicates the proportion of indigenous people within the populations of the arctic portions of the countries. The numbers are the total arctic populations of each country in the early 1990s. Indigenous people make up roughly 10% of the current population of the Arctic, though in the Canadian Arctic, they represent about half the population, and in Greenland, they are the majority.

- Saami Council (SC)
- Russian Association of Indigenous Peoples of the North (RAIPON)
- Aleut International Association (AIA)
- Inuit Circumpolar Conference (ICC)
- Gwich'in Council International (GCI)
- Arctic Athabaskan Council (AAC)

Executive Summary

Arctic Climate Change and Its Impacts

"Changes in climate that have already taken place are manifested in the decrease in extent and thickness of Arctic sea ice, permafrost thawing, coastal erosion, changes in ice sheets and ice shelves, and altered distribution and abundance of species."

IPCC, 2001

Earth's climate is changing, with the global temperature now rising at a rate unprecedented in the experience of modern human society. While some historical changes in climate have resulted from natural causes and variations, the strength of the trends and the patterns of change that have emerged in recent decades indicate that human influences, resulting primarily from increased emissions of carbon dioxide and other greenhouse gases, have now become the dominant factor.

These climate changes are being experienced particularly intensely in the Arctic. Arctic average temperature has risen at almost twice the rate as the rest of the world in the past few decades. Widespread melting of glaciers and sea ice and rising permafrost temperatures present additional evidence of strong arctic warming. These changes in the Arctic provide an early indication of the environmental and societal significance of global warming.

An acceleration of these climatic trends is projected to occur during this century, due to ongoing increases in concentrations of greenhouse gases in the earth's atmosphere. While greenhouse gas emissions do not primarily originate in the Arctic, they are projected to bring wide-ranging changes and impacts to the Arctic. These arctic changes will, in turn, impact the planet as a whole. For this reason, people outside the Arctic have a great stake in what is happening there. For example, climatic processes unique to the Arctic have significant effects on global and regional climate. The Arctic also provides important natural resources to the rest of the world (such as oil, gas, and fish) that will be affected by climate change. And melting of arctic glaciers is one of the factors contributing to sea-level rise around the globe.

Climate change is also projected to result in major impacts inside the Arctic, some of which are already underway. Whether a particular impact is perceived as negative or positive often depends on one's interests. For example, the reduction in sea ice is very likely to have devastating consequences for polar bears, ice-dependent seals, and local people for whom these animals are a primary food source. On the other hand, reduced sea ice is likely to increase marine access to the region's resources, expanding opportunities for shipping and possibly for offshore oil extraction (although operations could be hampered initially by increasing movement of ice in some areas). Further complicating the issue, possible increases in environmental damage that often accompanies shipping and resource extraction could harm the marine habitat and negatively affect the health and traditional lifestyles of indigenous people.

Another example is that increased areas of tree growth in the Arctic could serve to take up carbon dioxide and supply more wood products and related employment, providing local and global economic benefits.

On the other hand, increased tree growth is likely to add to regional warming and encroach on the habitat for many birds, reindeer/caribou, and other locally beneficial species, thereby adversely affecting local residents. Potential complications include projected increases in forest disturbances such as fires and insect outbreaks that could reduce expected benefits.

Unless we change our direction, we are likely to end up where we are headed.

Climate change is taking place within the context of many other ongoing changes in the Arctic, including the observed increase in chemical contaminants entering the Arctic from other regions, overfishing, land use changes that result in habitat destruction and fragmentation, rapid growth in the human population, and cultural, governance, and economic changes. Impacts on the environment and society result not from climate change alone, but from the interplay of all of these changes. This assessment has made an initial attempt to reveal some of this complexity, but limitations in current knowledge do not allow for a full analysis of all the interactions and their impacts.

One of the additional stresses in the Arctic that is addressed in this assessment results from increasing levels of ultraviolet radiation reaching the earth's surface due to stratospheric ozone depletion. As with many of the other stresses mentioned, there are important interactions between climate change and ozone depletion. The effects of climate change on the upper atmosphere make continued ozone depletion over the Arctic likely to persist for at least a few more decades. Thus, ultraviolet radiation levels in the Arctic are likely to remain elevated, and this will be most pronounced in the spring, when ecosystems are most sensitive to harmful ultraviolet radiation. The combination of climate change, excess ultraviolet radiation, and other stresses presents a range of potential problems for human health and well-being as well as risks to other arctic species and ecosystems.

The impacts of climate change in the Arctic addressed in this assessment are largely caused from outside the region, and will reverberate back to the global community in a variety of ways. The scientific findings reported here can inform decisions about actions to reduce the risks of climate change. As the pace and extent of climate change and its impacts increase, it will become more and more important for people everywhere to become aware of the changes taking place in the Arctic, and to consider them in evaluating what actions should be taken to respond.

Are These Impacts Inevitable?

Carbon dioxide concentrations in the atmosphere, which have risen rapidly due to human activities, will remain elevated above natural levels for centuries, even if emissions were to cease immediately. Some continued warming is thus inevitable. However, the speed and amount of warming can be reduced if future emissions are limited sufficiently to stabilize the concentrations of greenhouse gases. The scenarios developed by the IPCC assume a variety of different societal developments, resulting in various plausible levels of future emissions. None of these scenarios assume implementation of explicit policies to reduce greenhouse gas emissions. Thus, atmospheric concentrations do not level off in these scenarios, but rather continue to rise, resulting in significant increases in temperature and sea level and widespread changes in precipitation. The costs and difficulties of adapting to such increases are very likely to increase significantly over time.

If, on the other hand, society chooses to reduce emissions substantially, the induced changes in climate would be smaller and would happen more slowly. This would not eliminate all impacts, especially some of the irreversible impacts affecting particular species. However, it would allow ecosystems and human societies as a whole to adapt more readily, reducing overall impacts and costs. The impacts addressed in this assessment assume continued growth in greenhouse gas emissions. Although it will be very difficult to limit near-term consequences resulting from past emissions, many longer-term impacts could be reduced significantly by reducing global emissions over the course of this century. This assessment did not analyze strategies for achieving such reductions, which are the subject of efforts by other bodies.

Key Findings

The Arctic is extremely vulnerable to observed and projected climate change and its impacts. The Arctic is now experiencing some of the most rapid and severe climate change on earth. Over the next 100 years, climate change is expected to accelerate, contributing to major physical, ecological, social, and economic changes, many of which have already begun. Changes in arctic climate will also affect the rest of the world through increased global warming and rising sea levels.

1. Arctic climate is now warming rapidly and much larger changes are projected.

- Annual average arctic temperature has increased at almost twice the rate as that of the rest of the world over the past few decades, with some variations across the region.
- Additional evidence of arctic warming comes from widespread melting of glaciers and sea ice, and a shortening of the snow season.
- Increasing global concentrations of carbon dioxide and other greenhouse gases due to human activities, primarily fossil fuel burning, are projected to contribute to additional arctic warming of about 4-7°C over the next 100 years.
- Increasing precipitation, shorter and warmer winters, and substantial decreases in snow cover and ice cover are among the projected changes that are very likely to persist for centuries.
- Unexpected and even larger shifts and fluctuations in climate are also possible.

2. Arctic warming and its consequences have worldwide implications.

- Melting of highly reflective arctic snow and ice reveals darker land and ocean surfaces, increasing absorption of the sun's heat and further warming the planet.
- Increases in glacial melt and river runoff add more freshwater to the ocean, raising global sea level and possibly slowing the ocean circulation that brings heat from the tropics to the poles, affecting global and regional climate.
- Warming is very likely to alter the release and uptake of greenhouse gases from soils, vegetation, and coastal oceans.
- Impacts of arctic climate change will have implications for biodiversity around the world because migratory species depend on breeding and feeding grounds in the Arctic.

3. Arctic vegetation zones are very likely to shift, causing wide-ranging impacts.

- Treeline is expected to move northward and to higher elevations, with forests replacing a significant fraction of existing tundra, and tundra vegetation moving into polar deserts.
- More-productive vegetation is likely to increase carbon uptake, although reduced reflectivity of the land surface is likely to outweigh this, causing further warming.
- Disturbances such as insect outbreaks and forest fires are very likely to increase in frequency, severity, and duration, facilitating invasions by non-native species.
- Where suitable soils are present, agriculture will have the potential to expand northward due to a longer and warmer growing season.

4. Animal species' diversity, ranges, and distribution will change.

- Reductions in sea ice will drastically shrink marine habitat for polar bears, ice-inhabiting seals, and some seabirds, pushing some species toward extinction.
- Caribou/reindeer and other land animals are likely to be increasingly stressed as climate change alters their access to food sources, breeding grounds, and historic migration routes.
- Species ranges are projected to shift northward on both land and sea, bringing new species into the Arctic while severely limiting some species currently present.
- As new species move in, animal diseases that can be transmitted to humans, such as West Nile virus, are likely to pose increasing health risks.
- Some arctic marine fisheries, which are of global importance as well as providing major contributions to the region's economy, are likely to become more productive. Northern freshwater fisheries that are mainstays of local diets are likely to suffer.

5. Many coastal communities and facilities face increasing exposure to storms.

- Severe coastal erosion will be a growing problem as rising sea level and a reduction in sea ice allow higher waves and storm surges to reach the shore.
- Along some arctic coastlines, thawing permafrost weakens coastal lands, adding to their vulnerability.
- The risk of flooding in coastal wetlands is projected to increase, with impacts on society and natural ecosystems.
- In some cases, communities and industrial facilities in coastal zones are already threatened or being forced to relocate, while others face increasing risks and costs.

6. Reduced sea ice is very likely to increase marine transport and access to resources.

- The continuing reduction of sea ice is very likely to lengthen the navigation season and increase marine access to the Arctic's natural resources.
- Seasonal opening of the Northern Sea Route is likely to make trans-arctic shipping during summer feasible within several decades. Increasing ice movement in some channels of the Northwest Passage could initially make shipping more difficult.
- Reduced sea ice is likely to allow increased offshore extraction of oil and gas, although increasing ice movement could hinder some operations.
- Sovereignty, security, and safety issues, as well as social, cultural, and environmental concerns are likely to arise as marine access increases.

7. Thawing ground will disrupt transportation, buildings, and other infrastructure.

- Transportation and industry on land, including oil and gas extraction and forestry, will increasingly be disrupted by the shortening of the periods during which ice roads and tundra are frozen sufficiently to permit travel.
- As frozen ground thaws, many existing buildings, roads, pipelines, airports, and industrial facilities are likely to be destabilized, requiring substantial rebuilding, maintenance, and investment.
- Future development will require new design elements to account for ongoing warming that will add to construction and maintenance costs.
- Permafrost degradation will also impact natural ecosystems through collapsing of the ground surface, draining of lakes, wetland development, and toppling of trees in susceptible areas.

8. Indigenous communities are facing major economic and cultural impacts.

- Many Indigenous Peoples depend on hunting polar bear, walrus, seals, and caribou, herding reindeer, fishing, and gathering, not only for food and to support the local economy, but also as the basis for cultural and social identity.
- Changes in species' ranges and availability, access to these species, a perceived reduction in weather predictability, and travel safety in changing ice and weather conditions present serious challenges to human health and food security, and possibly even the survival of some cultures.
- Indigenous knowledge and observations provide an important source of information about climate change. This knowledge, consistent with complementary information from scientific research, indicates that substantial changes have already occurred.

9. Elevated ultraviolet radiation levels will affect people, plants, and animals.

- The stratospheric ozone layer over the Arctic is not expected to improve significantly for at least a few decades, largely due to the effect of greenhouse gases on stratospheric temperatures. Ultraviolet radiation (UV) in the Arctic is thus projected to remain elevated in the coming decades.
- As a result, the current generation of arctic young people is likely to receive a lifetime dose of UV that is about 30% higher than any prior generation. Increased UV is known to cause skin cancer, cataracts, and immune system disorders in humans.
- Elevated UV can disrupt photosynthesis in plants and have detrimental effects on the early life stages of fish and amphibians.
- Risks to some arctic ecosystems are likely as the largest increases in UV occur in spring, when sensitive species are most vulnerable, and warming-related declines in snow and ice cover increase exposure for living things normally protected by such cover.

10. Multiple influences interact to cause impacts to people and ecosystems.

- Changes in climate are occurring in the context of many other stresses including chemical pollution, overfishing, land use changes, habitat fragmentation, human population increases, and cultural and economic changes.
- These multiple stresses can combine to amplify impacts on human and ecosystem health and well-being. In many cases, the total impact is greater than the sum of its parts, such as the combined impacts of contaminants, excess ultraviolet radiation, and climatic warming.
- Unique circumstances in arctic sub-regions determine which are the most important stresses and how they interact.

Warming in the Arctic is causing changes in nearly every part of the physical climate system. Some of these changes are highlighted below and explored in further detail throughout this report.

Rising Temperatures

Temperatures have increased sharply in recent decades over most of the region, especially in winter. Winter increases in Alaska and western Canada have been around 3-4°C over the past half century. Larger increases are projected this century.

Increasing Precipitation

Arctic precipitation has increased by about 8% on average over the past century. Much of the increase has come as rain, with the largest increases in autumn and winter. Greater increases are projected for the next 100 years.

Rising River Flows

River discharge to the ocean has increased over much of the Arctic during the past few decades and spring peak river flows are occurring earlier. These changes are projected to accelerate.

Thawing Permafrost

Permafrost has warmed by up to 2°C in recent decades, and the depth of the layer that thaws each year is increasing in many areas. Permafrost's southern limit is projected to shift northward by several hundred kilometers during this century.

Declining Snow Cover

Snow cover extent has declined about 10% over the past 30 years. Additional decreases of 10-20% by the 2070s are projected, with the greatest declines in spring.

"The world can tell us everything we want to know. The only problem for the world is that it doesn't have a voice. But the world's indicators are there. They are always talking to us."

Quitsak Tarkiasuk
Ivujivik, Canada

Diminishing Lake and River Ice

Later freeze-up and earlier break-up of river and lake ice have combined to reduce the ice season by one to three weeks in some areas. The strongest trends are over North America and western Eurasia.

Melting Glaciers

Glaciers throughout the Arctic are melting. The especially rapid retreat of Alaskan glaciers represents about half of the estimated loss of mass by glaciers worldwide, and the largest contribution by glacial melt to rising sea level yet measured.

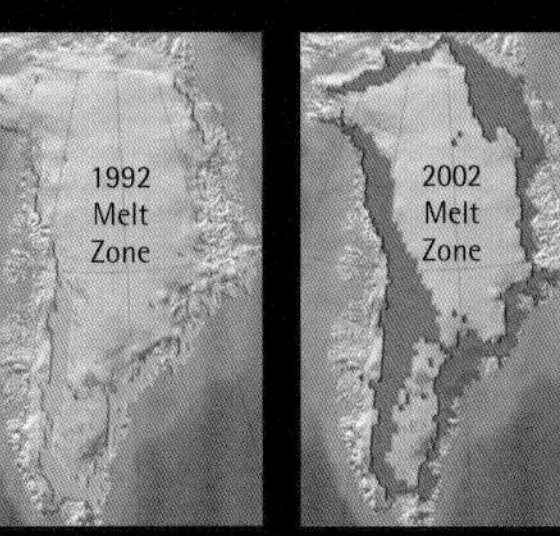

Melting Greenland Ice Sheet

The area of the Greenland Ice Sheet that experiences some melting has increased about 16% from 1979 to 2002. The area of melting in 2002 broke all previous records.

Retreating Summer Sea Ice

The average extent of sea-ice cover in summer has declined by 15-20% over the past 30 years. This decline is expected to accelerate, with the near total loss of sea ice in summer projected for late this century.

Rising Sea Level

Global and arctic sea level has risen 10-20 centimeters in the past 100 years. About an additional half meter of sea-level rise (with a range of 10 to 90 cm) is projected to occur during this century. The increase in the Arctic is projected to be greater than the global average. The slope of the land and whether the coastline is rising or falling also affects the relative sea-level rise in each location.

Less salty
Waters
1995 - 2000

Ocean Salinity Change

Reduced salinity and density have been observed in the North Atlantic Ocean as melting ice and increasing river runoff have added more freshwater to the ocean. If this trend persists, it could cause changes in ocean circulation patterns that strongly affect regional climate.

Impacts on Natural Systems

The climate trends highlighted on the previous pages affect natural ecosystems. Some of these impacts are highlighted below and explored throughout this report.

Wetland Changes

Permafrost thawing will cause lakes and wetlands to drain in some areas, while creating new wetlands in other places. The balance of these changes is not known, but as freshwater habitats are thus modified, major species shifts are likely.

Vegetation Shifts

Vegetation zones are projected to shift northward, with forests encroaching on tundra, and tundra encroaching on polar deserts. Limitations in amount and quality of soils are likely to slow this transition in some areas.

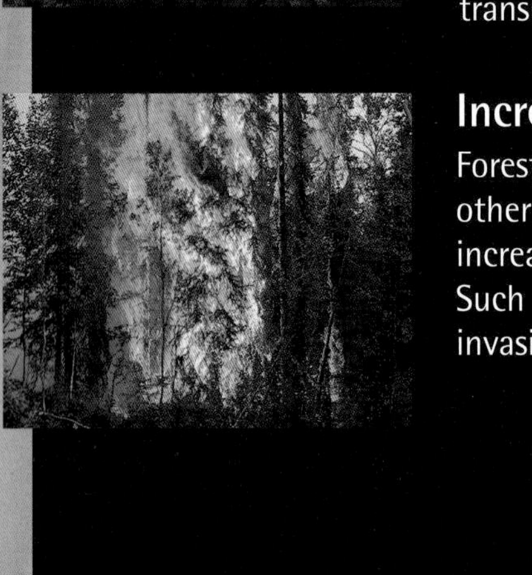

Increasing Fires and Insects

Forest fires, insect infestations, and other disturbances are projected to increase in frequency and intensity. Such events can subject habitats to invasion by non-native species.

Northward Species Shifts

The ranges of many plant and animal species are projected to shift northward, resulting in an increased number of species in the Arctic. Some currently widespread arctic species are likely to suffer major declines.

Marine Species at Risk

Marine species dependent on sea ice, including polar bears, ice-living seals, walrus, and some marine birds, are very likely to decline, with some species facing extinction.

Land Species at Risk

Species quite specifically adapted to the arctic climate are especially at risk including many species of mosses and lichens, lemmings, voles, arctic fox, and snowy owl.

UV Impacts

Increased ultraviolet radiation reaching the earth's surface as a result of stratospheric ozone depletion and the reduction in spring snow and ice cover will impact ecosystems on land and in water.

Old-growth Forest Loss

Old-growth forest is rich in species of lichens, mosses, fungi, insects, woodpeckers, and birds that nest in tree cavities. Climate warming would increase forest fires and insect-caused tree death, further reducing this valuable habitat which is already declining due to other human activities.

Carbon Cycle Changes

Over time, replacement of arctic vegetation with more productive vegetation from the south is likely to increase carbon dioxide uptake. On the other hand, methane emissions, mainly from warming wetlands and thawing permafrost, are likely to increase.

"Climate change in polar regions is expected to be among the largest and most rapid of any region on the Earth, and will cause major physical, ecological, sociological, and economic impacts, especially in the Arctic..."

IPCC, 2001

Projected permafrost boundary
Present tree-line
Projected tree-line
Projected summer sea-ice extent
Present permafrost boundary
Present Summer sea-ice extent

Changes in summer sea-ice extent and tree-line are projected to occur by the end of this century. The change in the permafrost boundary assumes that present areas of discontinuous permafrost will be free of any permafrost in the future and this is likely to occur beyond the 21st century.

Impacts on Society

The changes in climate and natural systems highlighted on the previous pages are projected to lead to numerous impacts on society throughout the Arctic.

Loss of Hunting Culture

For Inuit, warming is likely to disrupt or even destroy their hunting and food-sharing culture as reduced sea ice causes the animals on which they depend to decline, become less accessible, and possibly become extinct.

Declining Food Security

Access to traditional foods including seal, polar bear, caribou, and some fish and bird species is likely to be seriously impaired by climate warming. Reduced quality of food sources, such as diseased fish and dried up berries, are already being observed in some locations. Shifting to a more Western diet carries risks of increased diabetes, obesity, and cardiovascular diseases.

Human Health Concerns

Human health concerns also include increased accident rates due to environmental changes such as sea ice thinning, and health problems caused by adverse impacts on sanitation infrastructure due to thawing permafrost.

Wildlife Herd Impacts

Caribou and reindeer herds will face a variety of climate-related changes in their migration routes, calving grounds, and forage availability as snow and river ice conditions change, thus affecting the people who depend on hunting and herding them.

Expanding Marine Shipping

Shipping through key marine routes, including the Northern Sea Route and the Northwest Passage, is likely to increase. The summer navigation season is projected to lengthen considerably as the century progresses, due to the decline of sea ice. Expansion of tourism and marine transport of goods are likely outcomes.

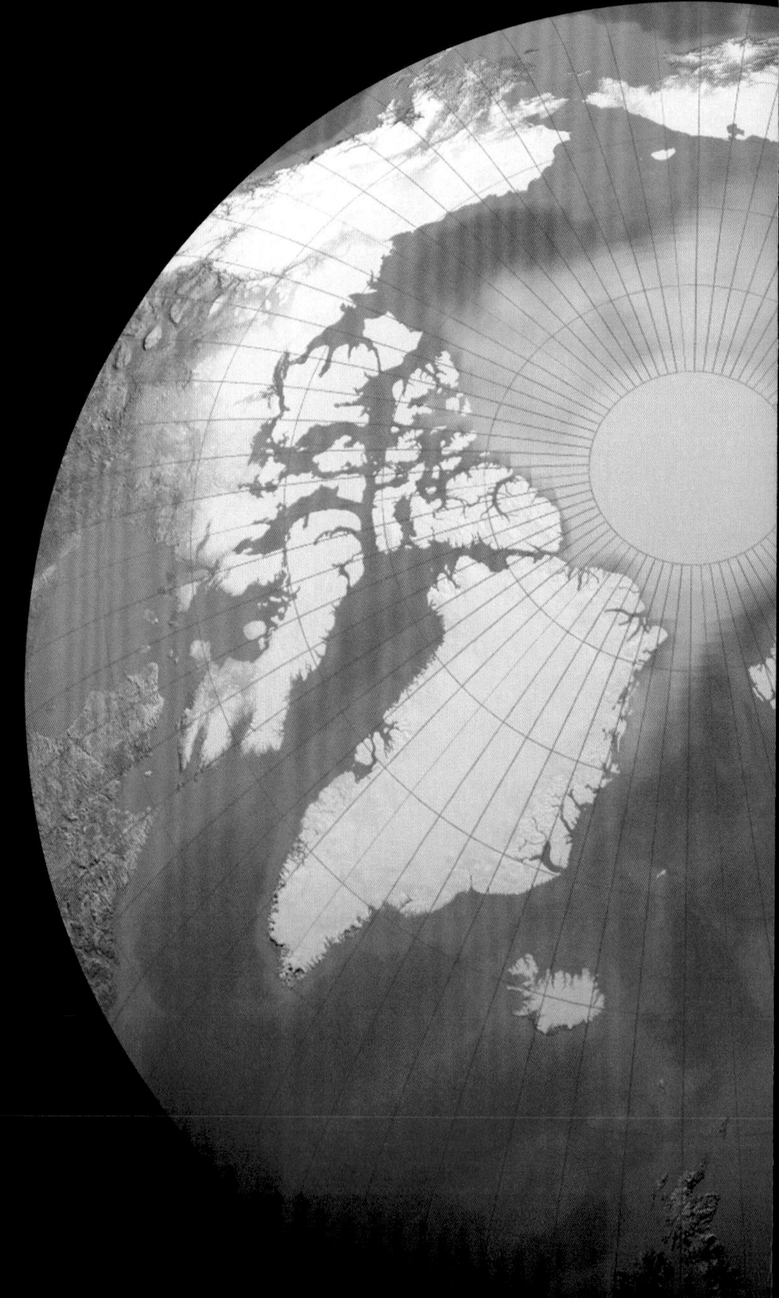

Increasing Access to Resources

Marine access to some arctic resources, including offshore oil and gas and some minerals, is likely to be enhanced by the reduction in sea ice, bringing new opportunities as well as environmental concerns. Increased ice movement could initially make some operations more difficult.

Enhanced Marine Fisheries

Some major arctic marine fisheries, including those for herring and cod, are likely to become more productive as climate warms. Ranges and migration patterns of many fish species are very likely to change.

Nowadays snows melt earlier in the springtime. Lakes, rivers, and bogs freeze much later in the autumn. Reindeer herding becomes more difficult as the ice is weak and may give way... Nowadays the winters are much warmer than they used to be. Occasionally during wintertime it rains. We never expected this; we could not be ready for this. It is very strange... The cycle of the yearly calendar has been disturbed greatly and this affects the reindeer herding negatively for sure."

Larisa Avdeyeva

Lovozero, Russia

Disrupted Transport on Land

Transportation routes and pipelines on land are already being disturbed in some places by thawing ground, and this problem is likely to expand. Oil and gas extraction and forestry will be increasingly disrupted by the shrinking of the period during which ice roads and tundra are sufficiently frozen to allow industrial operations. Northern communities that rely on frozen roadways to truck in supplies are also being affected.

Decline in Northern Freshwater Fisheries

Decreased abundance and local and global extinctions of arctic-adapted fish species are projected for this century. Arctic char, broad whitefish, and Arctic cisco, which are major contributors to the diets of local people, are among the species threatened by a warming climate.

Enhanced Agriculture and Forestry

Agricultural and forestry opportunities are likely to increase as potential areas for food and wood production expand northward due to a longer and warmer growing season and increasing

Sub-Regional Overview

In a region as large and diverse as the Arctic, there are significant sub-regional variations in climate. Recent warming has been more dramatic in some areas than others. A few places, such as parts of Canada and Greenland surrounding the Labrador Sea, have not yet experienced the widespread warming of the rest of the region, and have actually cooled. Regional variations in future climate change are also projected. Local features of the natural world and societies also create differences in what impacts will occur and which will be most significant in each sub-region.

For this assessment, four sub-regions were identified, and this report highlights selected impacts in each of these sub-regions. This is not a comprehensive evaluation of climate change impacts in these areas, nor an appraisal of which impacts are the most significant. Rather, it is a brief selection of important examples that emerged from this assessment. Further details can be found on pages 114-121 of this report. Some impacts are important in all of the sub-regions, but to avoid repetition, are not specifically discussed in each. Other assessments, some already underway, will examine the impacts of some specific activities, such as oil extraction, in these Arctic sub-regions.

In assessing future impacts in the sub-regions, projected changes in climate were primarily derived from global scale climate models. As regional scale climate models improve and become more widely available, future assessments may be capable of more precisely detailing the local and regional patterns of change. For this assessment, the patterns of climate change and their impacts should be viewed at a fairly broad regional scale, as they become less certain and less specific at smaller scales.

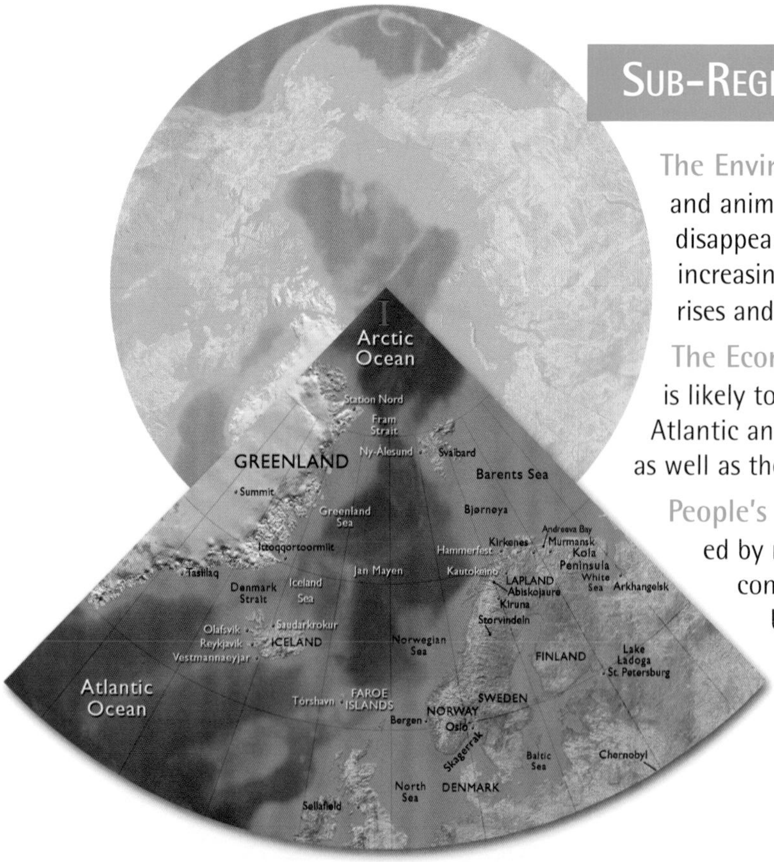

Sub-Region I — East Greenland, Iceland, Norway, Sweden, Finland, Northwest Russia, and adjacent seas

The Environment Northward shifts in the ranges of plant and animal species are very likely, with some tundra areas disappearing from the mainland. Low-lying coastal areas are increasingly likely to be inundated by storm surges as sea level rises and sea ice retreats.

The Economy Marine access to oil, gas, and mineral resources is likely to improve as sea ice retreats. A general increase in North Atlantic and Arctic fisheries is likely, based on traditional species as well as the influx of more southerly species.

People's Lives Reindeer herding is likely to be adversely affected by reduced snow cover and changing snow conditions. Traditional harvests of animals are likely to become more risky and less predictable. Animal diseases that can be transmitted to humans are likely to emerge.

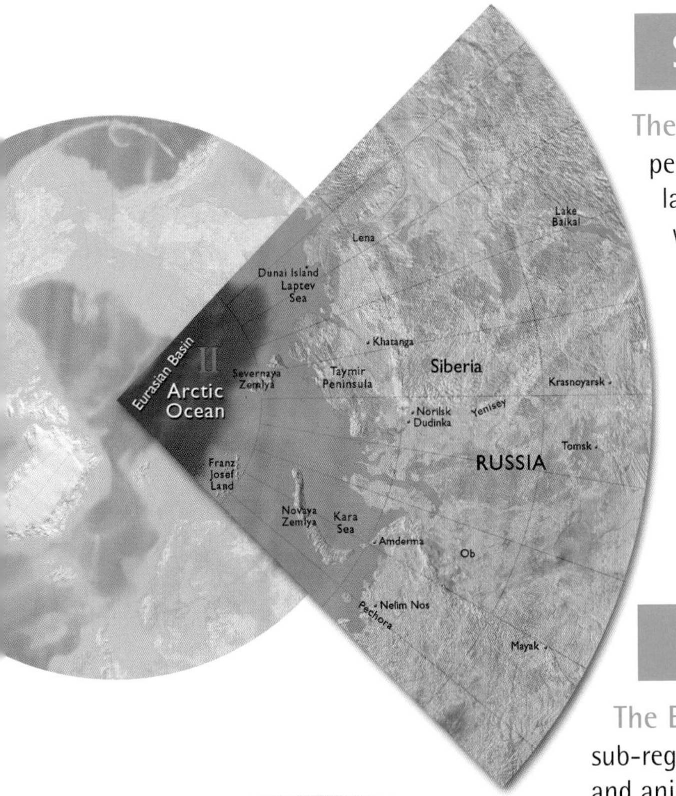

Sub-Region II Siberia and adjacent seas

The Environment Forests are likely to change significantly as climate warms, permafrost thaws, and fire and insect disturbances increase. Forests and shrublands are very likely to replace tundra in many areas. Plant and animal species will shift northward. River discharge will increase.

The Economy Sea-ice retreat is very likely to increase the navigation season through the Northern Sea Route, presenting economic opportunities as well as pollution risks. Access to offshore oil and gas is likely to improve but some activities could be hindered by increased wave action.

People's Lives Permafrost thawing is already causing serious damage to buildings and industrial facilities and is projected to continue. A shrinking river ice season and thawing permafrost are likely to hinder reindeer migration routes, affecting traditional livelihoods of indigenous people.

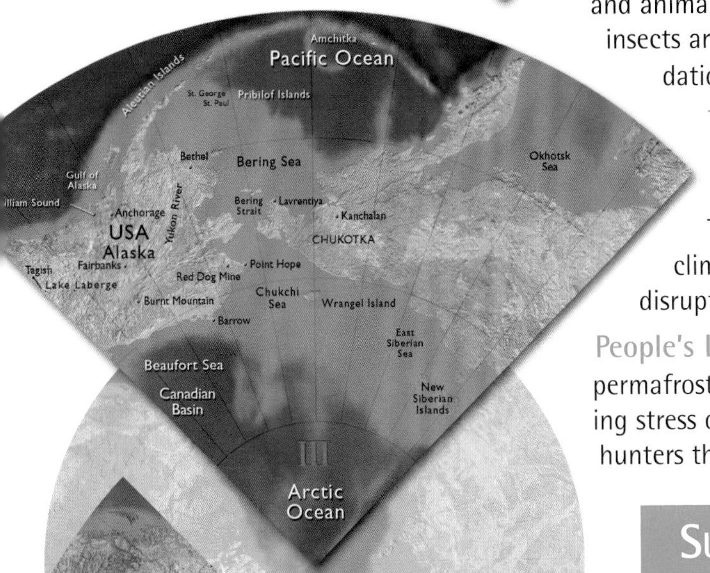

Sub-Region III Chukotka, Alaska, Western Canadian Arctic, and adjacent seas

The Environment Biological diversity is most at risk from climate change in this sub-region because it is currently home to the highest number of threatened plant and animal species in the Arctic. Increasing forest disturbances due to fires and insects are projected. Low-lying coastal areas will experience more frequent inundation.

The Economy Damage to infrastructure will result from permafrost thawing and coastal erosion. Reduced sea ice will enhance ocean access to northern coastlines. Thawing will hinder land transport in winter. Traditional local economies based on resources that are vulnerable to climate change (such as polar bears and ringed seals), are very likely to be disrupted by warming.

People's Lives Coastal erosion due to sea-ice decline, sea-level rise, and thawing permafrost is very likely to force the relocation of some villages and create increasing stress on others. Declines in ice-dependent species and increasing risks to hunters threaten the food security and traditional lifestyles of indigenous people.

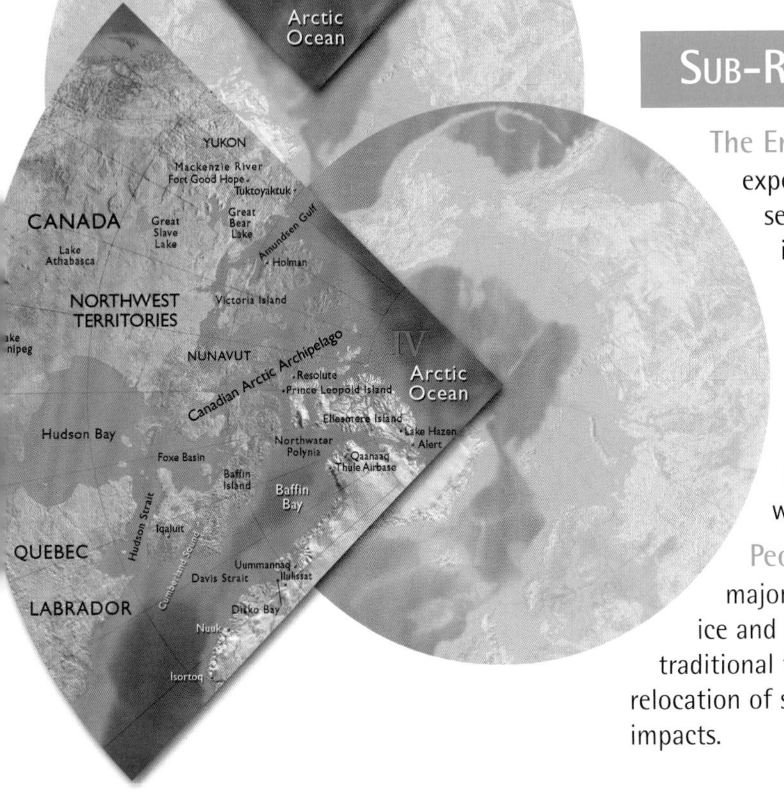

Sub-Region IV Central and Eastern Canadian Arctic, West Greenland, and adjacent seas

The Environment The Greenland Ice Sheet is likely to continue to experience record melting, changing the local environment and raising sea levels globally. Low-lying coastal areas will be more frequently inundated due to rising sea levels and storm surges.

The Economy Sea-ice retreat is likely to increase shipping through the Northwest Passage, providing economic opportunities while raising the risks of pollution due to oil spills and other accidents. More southerly marine fish species such as haddock, herring, and blue fin tuna could move into the region. Lake trout and other freshwater fish will decline, with impacts on local food supplies as well as sport fishing and tourism.

People's Lives Some Indigenous Peoples, particularly the Inuit, face major threats to their food security and hunting cultures as reduced sea ice and other warming-related changes hinder availability of and access to traditional food sources. Increases in sea level and storm surges could force the relocation of some low-lying coastal communities, causing substantial social impacts.

Why Does the Arctic Warm Faster than Lower Latitudes?

First, as arctic snow and ice melt, the darker land and ocean surfaces that are revealed absorb more of the sun's energy, increasing arctic warming. Second, in the Arctic, a greater fraction of the extra energy received at the surface due to increasing concentrations of greenhouse gases goes directly into warming the atmosphere, whereas in the tropics, a greater fraction goes into evaporation. Third, the depth of the atmospheric layer that has to warm in order to cause warming of near-surface air is much shallower in the Arctic than in the tropics, resulting in a larger arctic temperature increase. Fourth, as warming reduces the extent of sea ice, solar heat absorbed by the oceans in the summer is more easily transferred to the atmosphere in the winter, making the air temperature warmer than it would be otherwise. Finally, because heat is transported to the Arctic by the atmosphere and oceans, alterations in their circulation patterns can also increase arctic warming.

1. As snow and ice melt, darker land and ocean surfaces absorb more solar energy.

2. More of the extra trapped energy goes directly into warming rather than into evaporation.

3. The atmospheric layer that has to warm in order to warm the surface is shallower in the Arctic.

4. As sea ice retreats, solar heat absorbed by the oceans is more easily transferred to the atmosphere.

5. Alterations in atmospheric and oceanic circulation can increase warming.

Supporting Evidence for the Key Findings

This satellite image of an Ellesmere Island glacier that reaches the sea in the Greely Fjord reveals growing meltwater ponds on the glacier's surface as well as icebergs that have calved off the glacier and are floating in the fjord.

1 Arctic climate is now warming rapidly and much larger changes are projected.

Observed Changes in Climate

Records of increasing temperatures, melting glaciers, reductions in extent and thickness of sea ice, thawing permafrost, and rising sea level all provide strong evidence of recent warming in the Arctic. There are regional variations due to atmospheric winds and ocean currents, with some areas showing more warming than others and a few areas even showing a slight cooling; but for the Arctic as a whole, there is a clear warming trend. There are also patterns within this overall trend; for example, in most places, temperatures in winter are rising more rapidly than in summer. In Alaska and western Canada, winter temperatures have increased as much as 3-4°C in the past 50 years.

Observations suggest that precipitation has increased by roughly 8% across the Arctic over the past 100 years, although uncertainties in measuring precipitation in the cold arctic environment and the sparseness of data in parts of the region limit confidence in these results. There are regional variations in precipitation across the Arctic, and there will be regional variations in the changes in precipitation as well.

Arctic air temperatures are rising, with the strongest trends in winter and over the last few decades.

In addition to the overall increase, changes in the characteristics of precipitation have also been observed. Much of the precipitation increase appears to be coming as rain, mostly in winter, and to a lesser extent in autumn and spring. The increasing winter rains, which fall on top of existing snow, cause faster snowmelt and, when the rainfall is intense, can result in flash flooding in some areas. Rain-on-snow events have increased significantly across much of the Arctic, for example, by 50% over the past 50 years in western Russia.

In order to assess whether recent changes in arctic climate are unusual, that is, outside the range of natural variability, it is helpful to compare recent observations to records of how climate has behaved in the past. Data on past

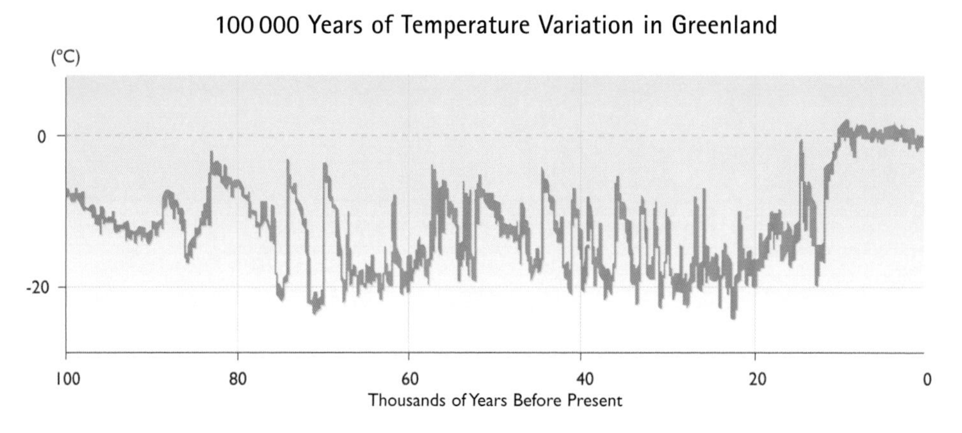

This record of temperature change (departures from present conditions) has been reconstructed from a Greenland ice core. The record demonstrates the high variability of the climate over the past 100 000 years. It also suggests that the climate of the past 10 000 years or so, which was the time during which human civilization developed, has been unusually stable. There is concern that the rapid warming caused by the increasing concentrations of greenhouse gases due to human activities could destabilize this state.

climate come from ice cores and other sources that provide reasonable representations of what climate was like in the distant past. Examining the record of past climatic conditions indicates that the amount, speed, and pattern of warming experienced in recent decades are indeed unusual and are characteristic of the human-caused increase in greenhouse gases.

Both natural and human-caused factors can influence the climate. Among the natural factors that can have significant effects lasting years to decades are variations in solar output, major volcanic eruptions, and natural, sometimes cyclic, interactions between the atmosphere and oceans. Several important natural modes of variability that especially affect the Arctic have been identified, including the Arctic Oscillation, the Pacific Decadal Oscillation, and the North Atlantic Oscillation. Each of these can affect the regional patterns of such features as the intensity and tracks of storm systems, the direction of the prevailing winds, the amount of snow, and the extent of sea ice. In addition to changing long-term average climatic conditions, human-induced changes in the climate may also affect the intensity, patterns, and features of these natural variations.

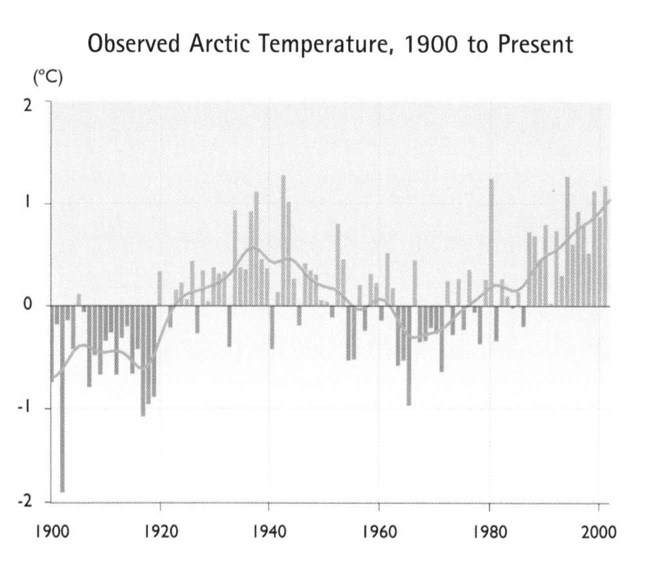

Annual average change in near surface air temperature from stations on land relative to the average for 1961-1990, for the region from 60 to 90°N.

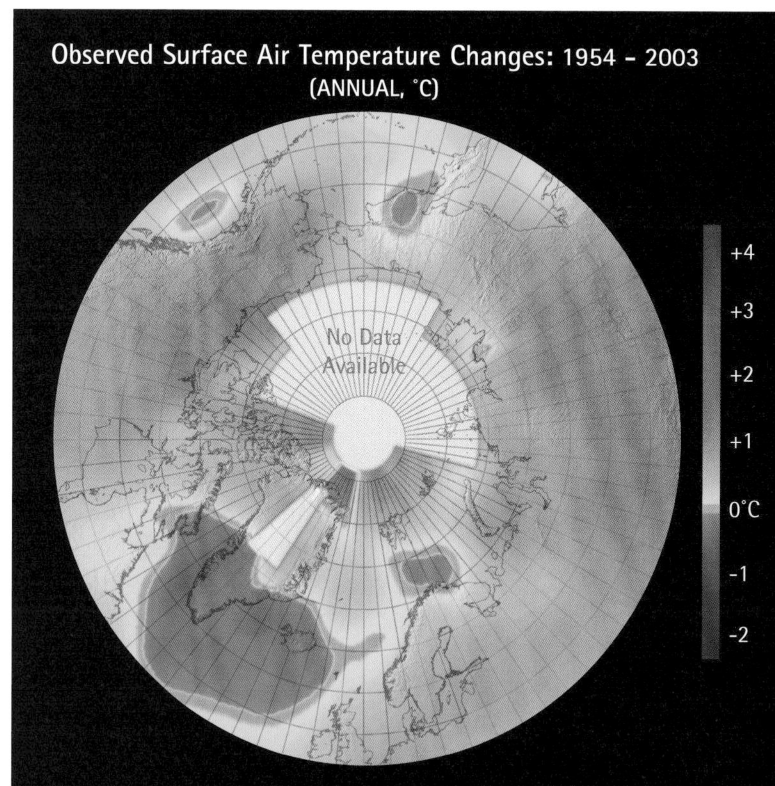

The colors indicate the change in temperature from 1954 to 2003. The map above indicates annual average temperature change, which ranges from a 2-3°C warming in Alaska and Siberia to a cooling of up to 1°C in southern Greenland.

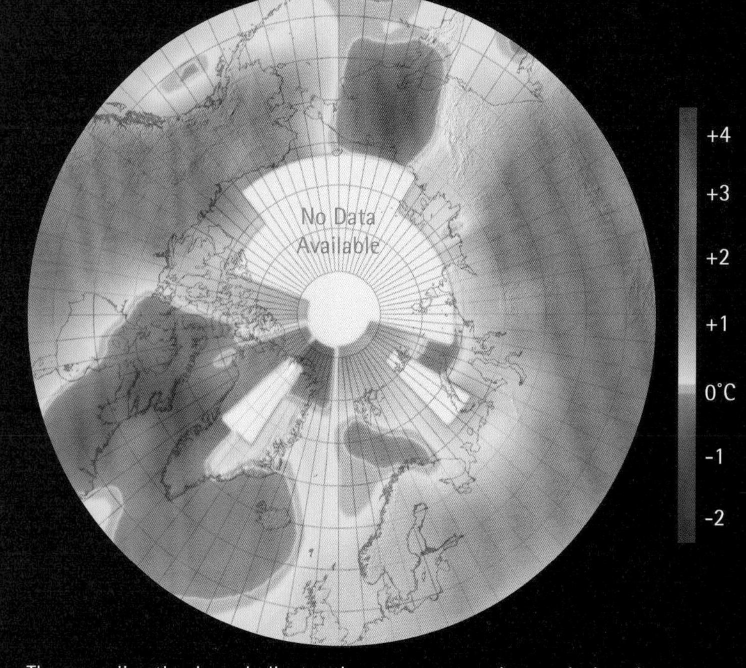

The map directly above indicates the temperature change during the winter months, ranging from a warming of up to 4°C in Siberia and Northwest Canada to a cooling of 1°C over southern Greenland.

1 Arctic climate is now warming rapidly and much larger changes are projected.

Changes in Sea Ice: A Key Climate Change Indicator

"Climate" refers to much more than just temperature and precipitation. In addition to long-term average weather conditions, climate also includes extreme events, as well as aspects of the system such as snow, ice, and circulation patterns in the atmosphere and oceans. In the Arctic, sea ice is one of the most important climatic variables. It is a key indicator and agent of climate change, affecting surface reflectivity, cloudiness, humidity, exchanges of heat and moisture at the ocean surface, and ocean currents. And as illustrated later in this report, changes in sea ice have enormous environmental, economic, and societal implications.

Just as miners once had canaries to warn of rising concentrations of noxious gases, researchers working on climate change rely on arctic sea ice as an early warning system. The sea ice presently covering the Arctic Ocean and neighboring seas is highly sensitive to temperature changes in the air above and the ocean below. Over recent decades, Arctic watchers detected a slow shrinkage of the ice pack, suggestive of the initial influences of global warming. In recent years, the rate of retreat has accelerated, indicating that the "canary" is in trouble.

"Ice is a supporter of life. It brings the sea animals from the north into our area and in the fall it also becomes an extension of our land. When it freezes along the shore, we go out on the ice to fish, to hunt marine mammals, and to travel... When it starts disintegrating and disappearing faster, it affects our lives dramatically."

Caleb Pungowiyi
Nome, Alaska

AN ICE PRIMER

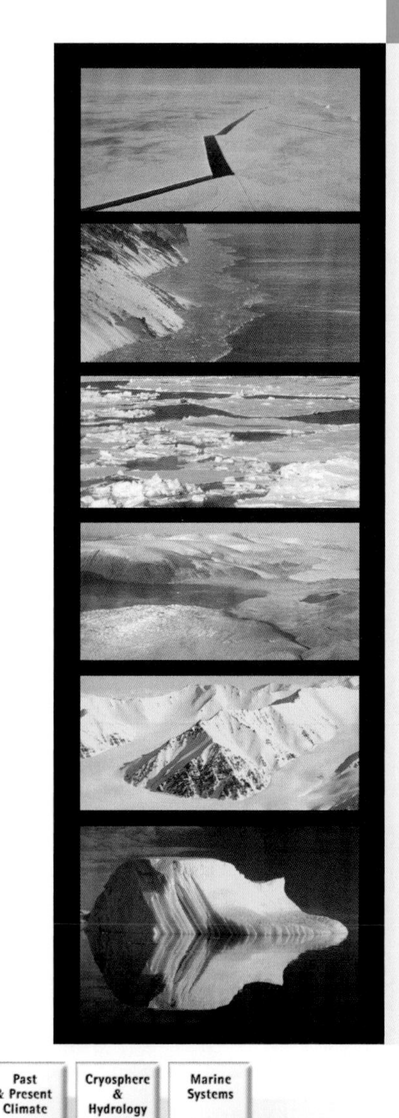

Sea ice is formed as seawater freezes. Because sea ice is less dense than seawater, it floats on top of the ocean. As sea ice forms, it rejects the majority of its salt to the ocean, making the ice even lighter. Because sea ice is formed from existing sea water, its melting does not raise sea level.

Fast ice is sea ice that grows from the coast into the sea, remaining attached to the coast or grounded to a shallow sea floor. It is important as a resting, hunting, and migration platform for species such as polar bears and walrus.

Pack ice refers to a large area of floating sea ice fragments that are packed together.

Ice caps and **glaciers** are land-based ice, with ice caps "capping" hills and mountains and glaciers usually referring to the ice filling the valleys, although the term glacier is often used to refer to ice caps as well.

An **ice sheet** is a collection of ice caps and glaciers, such as currently found on Greenland and on Antarctica. When ice caps, glaciers, and ice sheets melt, they cause sea level to rise by adding to the amount of water in the oceans.

An **iceberg** is a chunk of ice that calves off a glacier or ice sheet and floats at the ocean surface.

Over the past 30 years, the annual average sea-ice extent has decreased by about 8%, or nearly one million square kilometers, an area larger than all of Norway, Sweden, and Denmark combined, and the melting trend is accelerating. Sea-ice extent in summer has declined more dramatically than the annual average, with a loss of 15-20% of the late-summer ice coverage. There is also significant variability from year to year. September 2002 had the smallest extent of arctic sea-ice cover on record, and September 2003 was very nearly as low.

Just as miners once had canaries to warn of rising concentrations of noxious gases, researchers working on climate change rely on arctic sea ice as an early warning system.

Sea ice has also become thinner in recent decades, with arctic-wide average thickness reductions estimated at 10-15%, and with particular areas showing reductions of up to 40% between the 1960s and late 1990s. Impacts of a decline in sea ice are discussed throughout this report and include increased air temperature, decreased salinity of the ocean's surface layer, and increased coastal erosion.

Observed seasonal Arctic sea-ice extent (1900-2003)

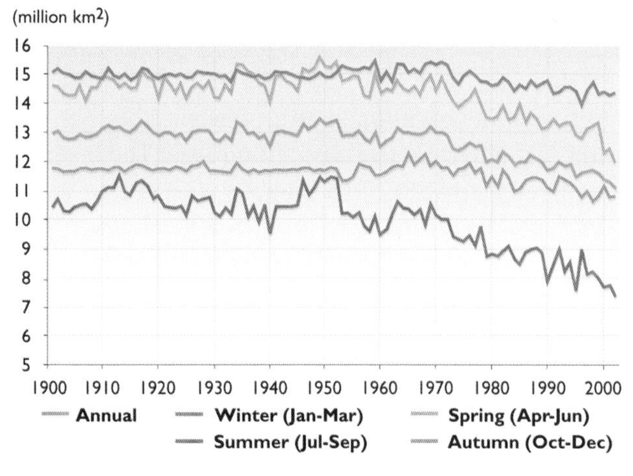

Annual average extent of arctic sea ice from 1900 to 2003. A decline in sea-ice extent began about 50 years ago and this decline sharpened in recent decades, corresponding with the arctic warming trend. The decrease in sea-ice extent during summer is the most dramatic of the trends.

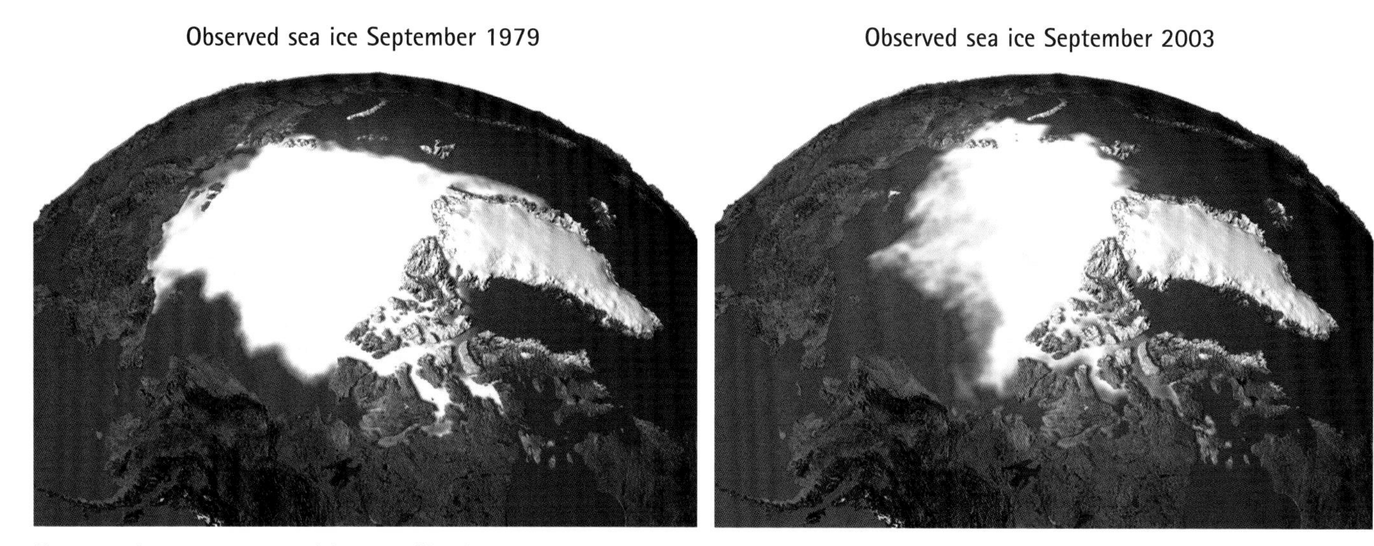

These two images, constructed from satellite data, compare arctic sea ice concentrations in September of 1979 and 2003. September is the month in which sea ice is at its yearly minimum and 1979 marks the first year that data of this kind became available in meaningful form. The lowest concentration of sea ice on record was in September 2002.

1 Arctic climate is now warming rapidly and much larger changes are projected.

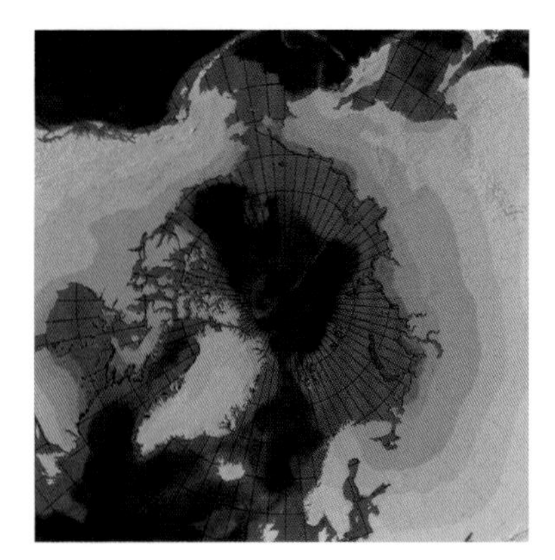

Projecting Future Climate

This assessment has drawn information from a variety of approaches for documenting past and present climatic conditions and projecting future climatic conditions, including: observed data (such as data from instruments like thermometers, and past climate records from tree rings, ice cores, and sediments); field experiments; computer-based climate models; and indigenous knowledge. When information from several methods converges, it offers greater confidence in the results. Still, there will always be uncertainties and surprises in projecting future changes in climate.

Projecting future climate change and its potential impacts is undertaken in a systematic manner. Two major factors determine how human activities will cause the climate to change in the future:

- the level of future global emissions of greenhouse gases, and
- the response of the climate system to these emissions.

Research over recent decades has contributed a great deal of information regarding each of these factors.

Even using the lowest emissions scenario, and the model that generates the least warming in response to changes in atmospheric composition, leads to a projection that the earth will warm more than twice as much in this century as it warmed over the past century.

Projecting the level of future emissions is carried out by developing plausible scenarios for future changes in population, economic growth, technological and political change, and other aspects of future human society that are difficult to fully anticipate. The IPCC produced a Special Report on Emissions Scenarios (SRES) to grapple with these issues. Their scenarios encompass a range of possible futures based on how societies, economies, and energy technologies are likely to evolve, and can be used to estimate the likely range of future emissions that affect the climate.

Regarding the response of the climate system, computer models developed by research centers from around the world represent aspects of the climate system (such as how clouds and ice cover might be expected to change, and ultimately how climate and sea level might be influenced) somewhat differently, resulting in differences in the degree of warming projected.

Projected Global Temperature Rise

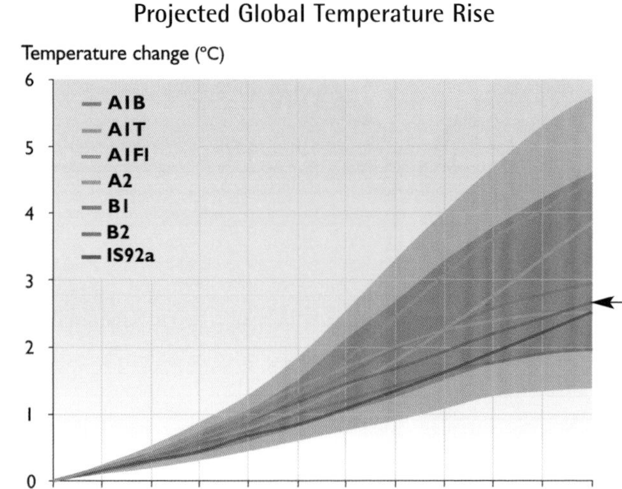

Projections of global temperature change (shown as departures from the 1990 temperature) from 1990 to 2100 for seven illustrative emissions scenarios. The brown line shows the projection of the B2 emissions scenario, the primary scenario used in this assessment, and the scenario on which the maps in this report showing projected climate changes are based. The pink line shows the A2 emissions scenario, used to a lesser degree in this assessment. The dark gray band shows the range of results for all the SRES emissions scenarios with one average model while the light gray band shows the full range of scenarios using various climate models.

Projected Arctic Temperature Rise

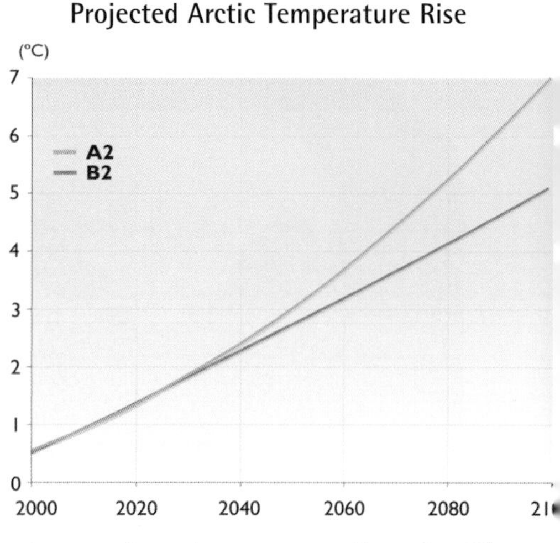

Increases in arctic temperature (for 60°-90°N) projected by an average of ACIA models for the A2 and B2 emissions scenarios, relative to 1981-2000.

Regardless of the emissions scenario or computer model selected, every model simulation projects significant global warming over the next 100 years. Even using the lowest emissions scenario, and the model that generates the least warming in response to changes in atmospheric composition, leads to a projection that the earth will warm more than twice as much in this century as it warmed over the past century. Model simulations further indicate that the warming in the Arctic will be substantially greater than the global average (in some regions, more than double). While the models differ in their projections of some of the features of climate change, they are all in agreement that the world will warm significantly as a result of human activities and that the Arctic is likely to experience noticeable warming particularly early and intensely.

The climate models and emissions scenarios reviewed by the IPCC generate a range of possible conditions for the future. To provide an indication of the types of impacts that are likely to occur, ACIA drew upon the results of five climate models from leading climate research centers and one moderate emissions scenario (known as B2, see Appendix 1 for more information) to be the primary basis for its assessment of the impacts of future climate conditions. The maps of projected climate conditions in this report are based on this B2 emissions scenario. A second emissions scenario (called A2) was added to some analyses to explore another possible future. The focus on these two scenarios here reflects a number of practical limits to conducting this assessment, and is not a judgment that these are the most likely outcomes.

When viewing the model results in this report, it is important to remember that these are not worst-case or best-case scenarios, but rather fall slightly below the middle of the range of temperature rise projected by global climate models. It is also important to note that for many of the impacts summarized in this report, information was also drawn from additional sources, including observed changes in climate, observed impacts, extrapolation of current trends, and laboratory and field experiments published in the peer-reviewed scientific literature.

When viewing the model results in this report, it is important to remember that these are not worst-case or best-case scenarios, but rather fall slightly below the middle of the range of temperature rise projected by global climate models.

Global Climate Model

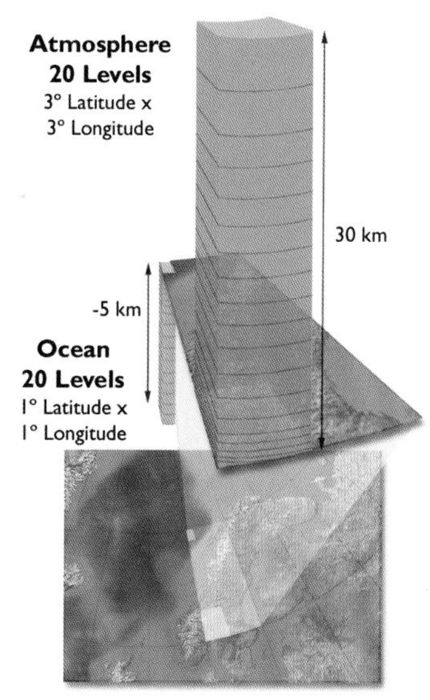

Global climate models are computer simulations based on physical laws represented by mathematical equations that are solved using a three-dimensional grid over the globe. The models include the major components of the climate system including the atmosphere, oceans, land surface, snow and ice, living things, and the processes that go on within and between them. As illustrated in the figure, the resolution (grid size) of the global models is fairly coarse, meaning that there is generally higher confidence in larger scale projections and greater uncertainty at increasingly small scales.

Projected Arctic Surface Air Temperatures 2000-2100

60°N - Pole: Change from 1981-2000 average

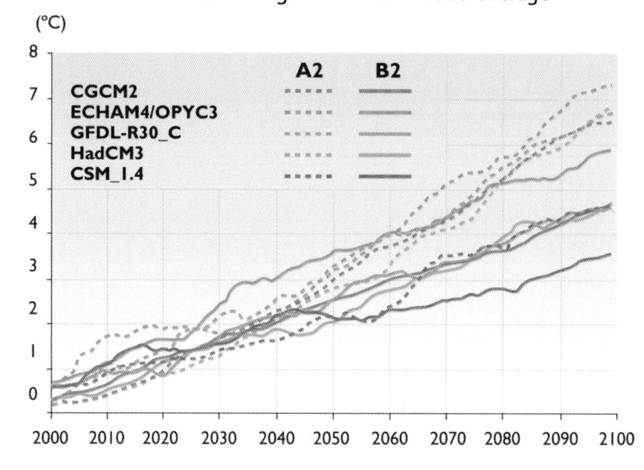

The ten lines show air temperatures for the region from 60°N to the pole as projected by each of the five ACIA global climate models using two different emissions scenarios. The projections remain similar through about 2040, showing about a 2°C temperature rise, but then diverge, showing increases from around 4° to over 7°C by 2100. The full range of models and scenarios reviewed by the IPCC cover a wider range of possible futures. Those used in this assessment fall roughly in the middle of this range, and thus represent neither best- nor worst-case scenarios.

Note: The full names of these models and a description of the A2 and B2 emissions scenarios can be found in Appendix 1, on pages 128 - 129.

1 Arctic climate is now warming rapidly and much larger changes are projected.

Projected Changes in Arctic Temperature

The maps below show projected changes in arctic temperature as an annual average, and for the winter (December, January, and February). They show the projected temperature change from the 1990s to the 2090s, based on the average change calculated by the five ACIA climate models using the B2 emissions scenario (resulting in a temperature rise slightly below the middle of the range of IPCC scenarios). Under this scenario, by the latter part of this century, annual average temperatures are projected to rise across the entire Arctic, with increases of roughly 3-5°C over the land areas and up to 7°C over the oceans. Winter temperatures are projected to rise significantly more, with increases of 4-7°C over the land areas and 7-10°C over the oceans. Some of the strongest warming is projected for land areas, such as northern Russia, which are adjacent to oceans in which sea ice is projected to decline sharply.

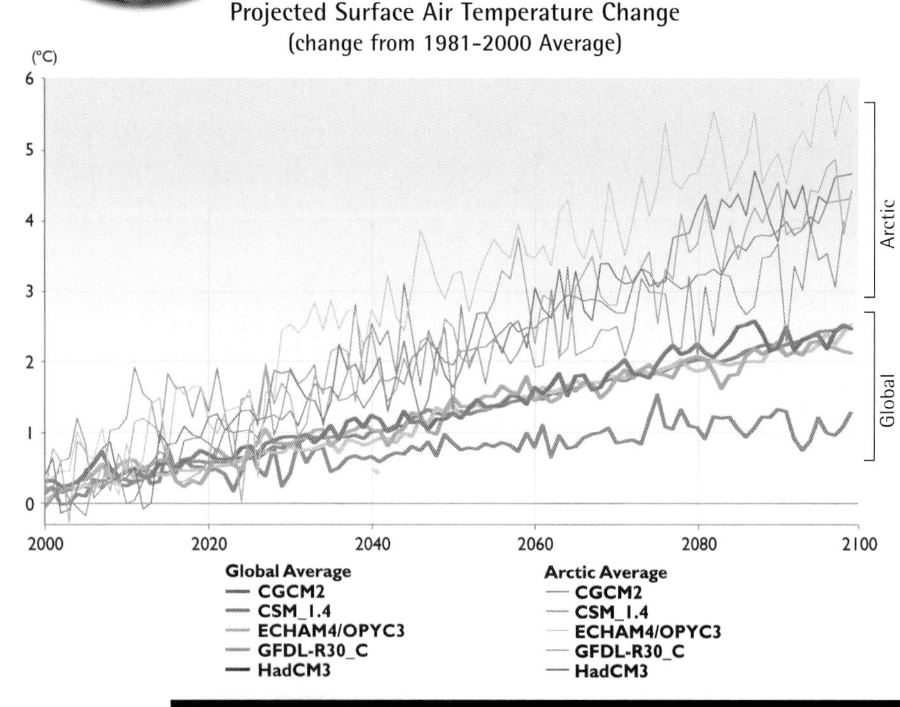

This graph shows average temperatures projected by the five ACIA climate models for the B2 emissions scenario. The heavy lines at the bottom are projected average **global** temperature increases and the thinner lines above are the projected **arctic** temperature increase. As the results show, the temperature increases are projected to be much greater in the Arctic than for the world as a whole. It is also apparent that the year-to-year variability is greater in the Arctic.

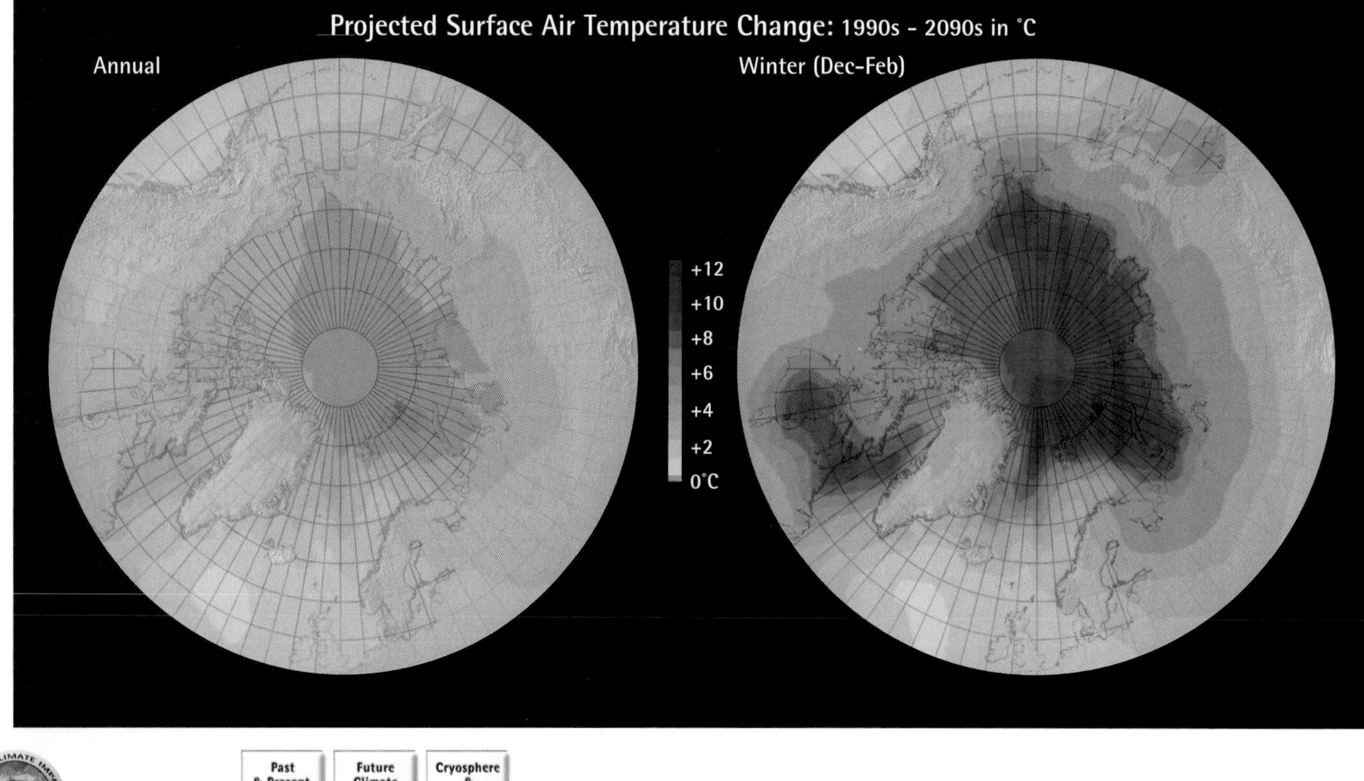

These maps show the projected temperature change from the 1990s to the 2090s, based on the average change calculated by the five ACIA climate models using the lower of the two emissions scenarios (B2) considered in this assessment. On these maps, orange indicates that an area is projected to warm by about 6°C from the 1990s to the 2090s.

Projected Changes in Arctic Precipitation

Global warming will lead to increased evaporation and in turn to increased precipitation (this is already occurring). Over the Arctic as a whole, annual total precipitation is projected to increase by roughly 20% by the end of this century, with most of the increase coming as rain. During the summer, precipitation over northern North America and Chukotka, Russia is projected to increase, while summer rainfall in Scandinavia is projected to decrease. During winter, precipitation for virtually all land areas (except southern Greenland) is projected to increase. The increase in arctic precipitation is projected to be most concentrated over coastal regions and in the winter and autumn; increases during these seasons are projected to exceed 30%.

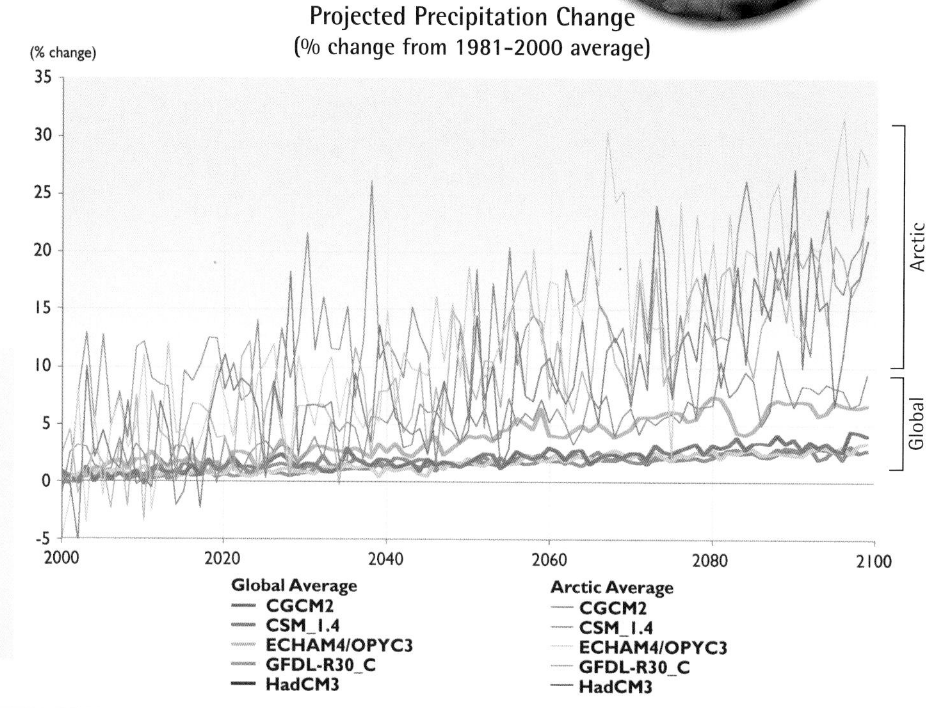

This graph shows percentage changes in average precipitation projected by the five ACIA climate models for the B2 emissions scenario. The heavy lines at the bottom are projected average **global** precipitation changes and the thinner lines above are projected **arctic** precipitation changes. As the results show, the precipitation increases are projected to be much greater in the Arctic than for the world as a whole. It is also apparent that the year-to-year variability is much greater in the Arctic.

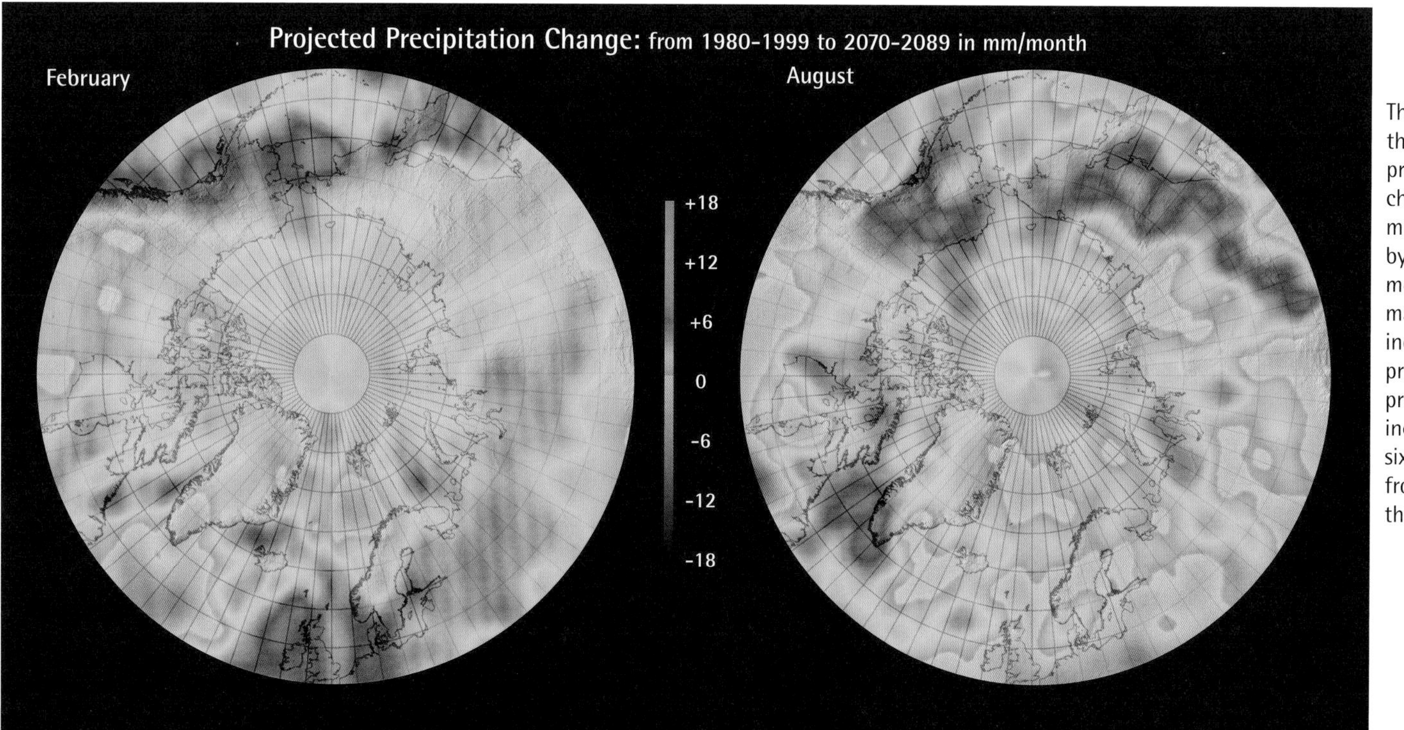

These maps show the projected precipitation change in mm per month, calculated by the ACIA climate models. On these maps, dark green indicates that precipitation is projected to increase by about six mm per month from the 1990s to the 2090s.

1 Arctic climate is now warming rapidly and much larger changes are projected.

The projected reductions in sea ice will increase regional and global warming by reducing the reflectivity of the ocean surface.

Projected Changes in Sea Ice

As noted earlier, sea ice has already declined considerably over the past half century. Additional declines of roughly 10-50% in annual average sea-ice extent are projected by 2100. Loss of sea ice during summer is projected to be considerably greater than the annual average decrease, with a 5-model average projecting more than a 50% decline by the end of this century, and some models showing near-complete disappearance of summer sea ice. The projected reductions in sea ice will increase regional and global warming by reducing the reflectivity of the ocean surface. Additional impacts of the projected sea-ice decline on natural systems and communities in the Arctic and around the world are discussed throughout this report.

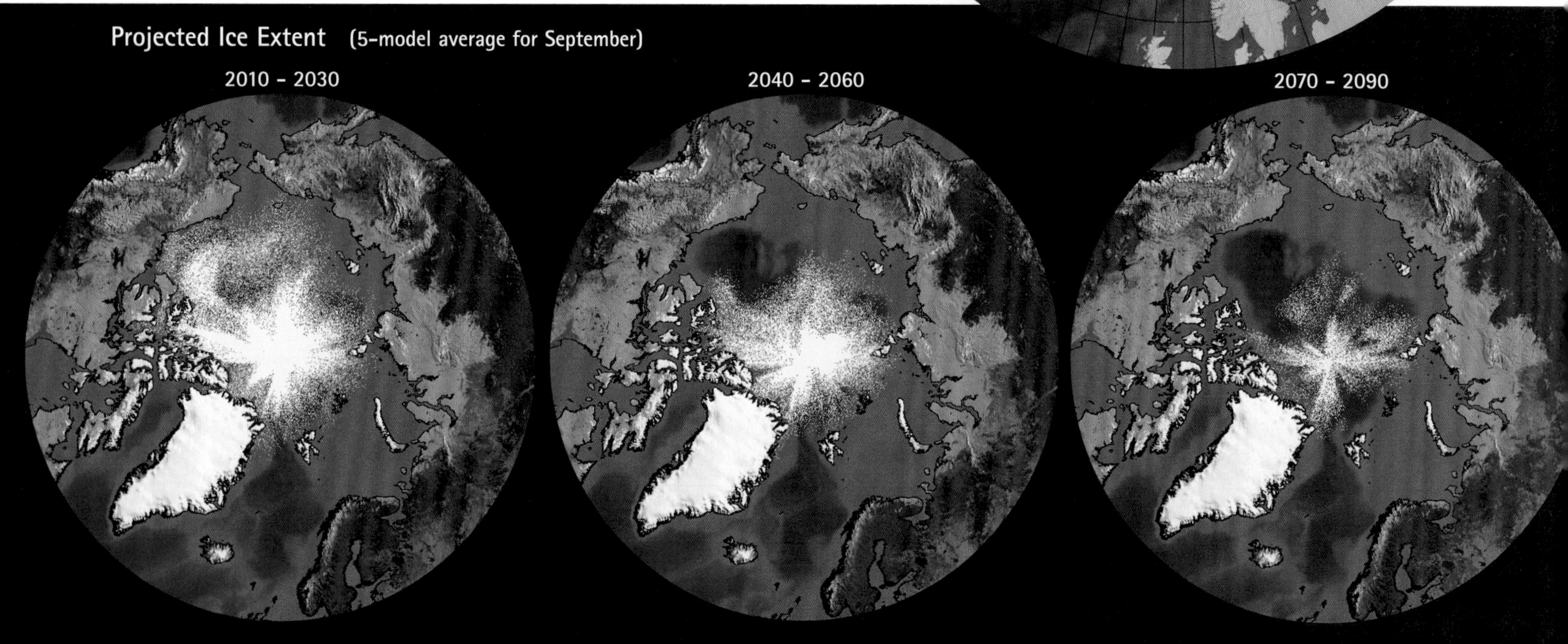

September sea-ice extent, already declining markedly, is projected to decline even more rapidly in the future. The three images above show the average of the projections from five climate models for three future time periods. As the century progresses, sea ice moves further and further from the coasts of arctic land masses, retreating to the central Arctic Ocean. Some models project the nearly complete loss of summer sea ice in this century.

Projected Changes in Snow Cover

Snow cover extent over arctic land areas has declined by about 10% over the past 30 years, and model projections suggest that it will decrease an additional 10-20% before the end of this century. The decreases in snow-covered area are projected to be greatest in spring (April and May), suggesting a further shortening of the snow season and an earlier pulse of river runoff to the Arctic Ocean and coastal seas. Important snow quality changes are also projected, such as an increase in thawing and freezing in winter that leads to ice layer formation that in turn restricts the access of land animals to food and nesting sites. Some impacts of the projected changes will include a reduction in the beneficial insulating effect of snow cover for vegetation and other living things and the ability of animals to forage. Flows of freshwater across the land to the ocean, and transfers of moisture and heat from the land to the atmosphere and marine systems will also be affected by the changes. Additional impacts of the snow cover decline are discussed throughout this report.

The decreases in snow-covered area are projected to be greatest in spring (April and May), suggesting a further shortening of the snow season and an earlier pulse of river runoff to the Arctic Ocean and coastal seas.

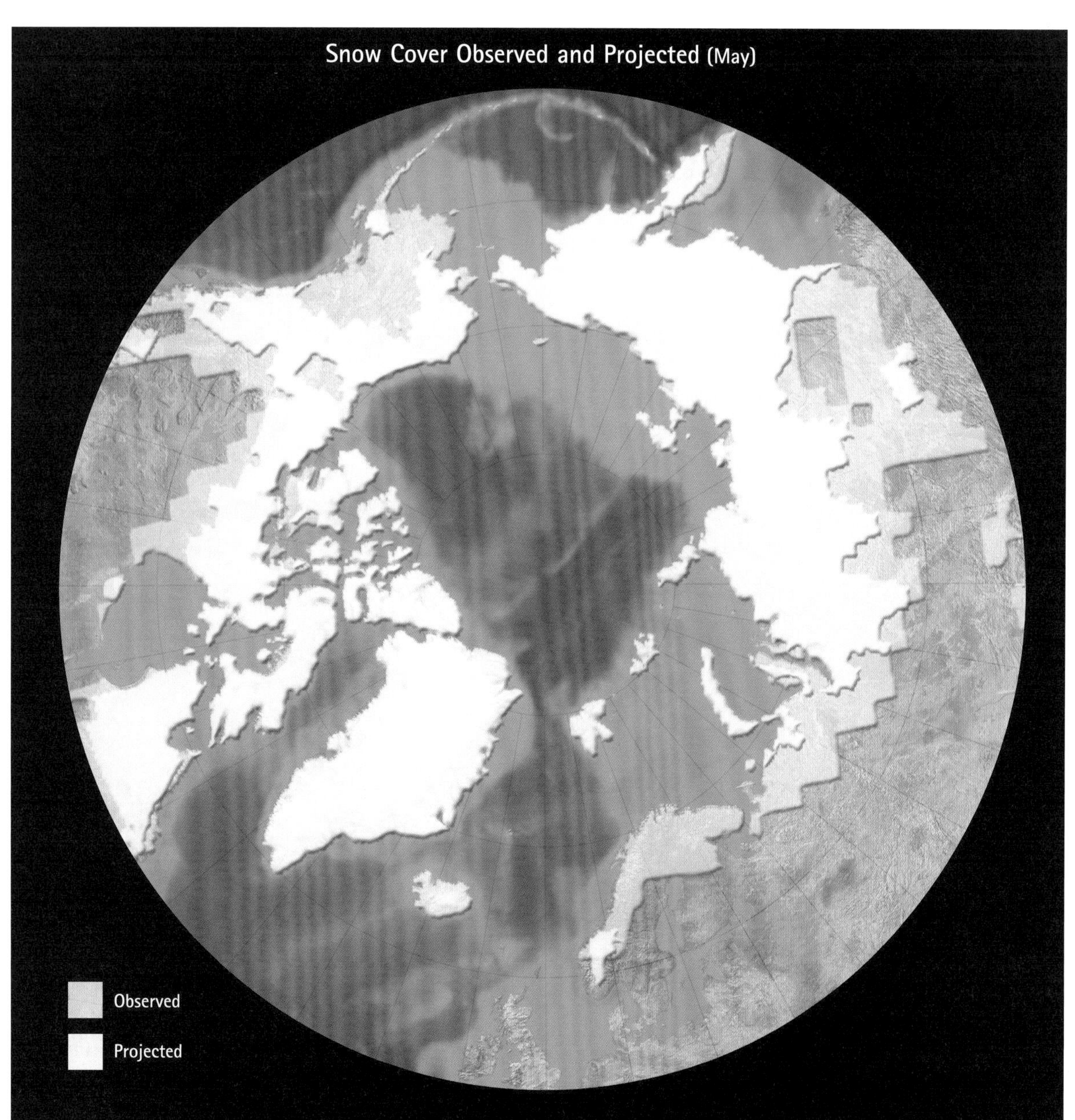

May snow cover is projected to decrease substantially throughout the Arctic. The gray area in the figure shows the current extent of May snow cover. The white area is the projected area of May snow cover in the 2070 to 2090 time period based on ACIA model projections. The large-scale pattern of projected snow cover retreat in spring is apparent.

1 Arctic climate is now warming rapidly and much larger changes are projected.

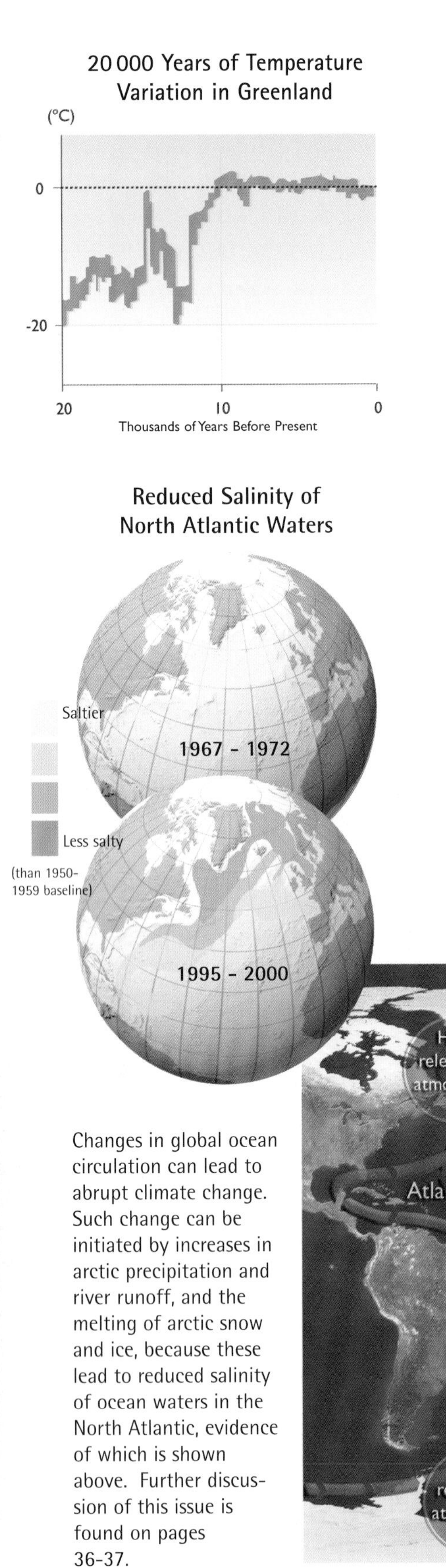

Abrupt Change

While most analyses of climate impacts in this and other assessments focus on scenarios of steady warming of the climate, there is also the possibility that gradual warming could trigger an abrupt change in climate. Such abrupt change could result from non-linear processes in the climate system. An example would be if a critical threshold (such as the freezing point) were crossed. Once a threshold is passed, the system could shift abruptly from one state to another. There is evidence that alternative stable states exist for components of the climate system, and for natural systems as well, though less is known about what triggers system shifts from one state to another. The mechanisms of abrupt change are thus not adequately represented in current climate models, leaving open the possibility of surprises. The idea that abrupt changes are plausible outcomes is also suggested by the relatively high natural variability of arctic climate as compared to conditions in the rest of the world. Records of past climate indicate, for example, that very large shifts in arctic climate patterns have apparently occurred over short timescales in the past.

For example, ice core records indicate that temperatures over Greenland dropped by as much as 5˚C within a few years during the period of warming that followed the last ice age, before abruptly warming again. This relatively sudden and then persistent change in the weather over Greenland was apparently driven by the crossing of a threshold involving ocean salinity that led to a sharp reduction in the ocean circulation that brings warmth to Europe and the Arctic. This oceanic change most likely prompted a shift in atmospheric circulation that lasted several centuries and caused large climatic changes over land areas surrounding the North Atlantic and beyond. Persistent, although smaller shifts in atmospheric circulation patterns (such as occur during the phase changes of the North Atlantic and Arctic Oscillations) occurred during the 20th century. These shifts apparently caused changes in the prevailing weather of arctic countries, contributing, for example, to warm decades, such as the 1930s and 1940s, and cool decades, such as the 1950s and 1960s.

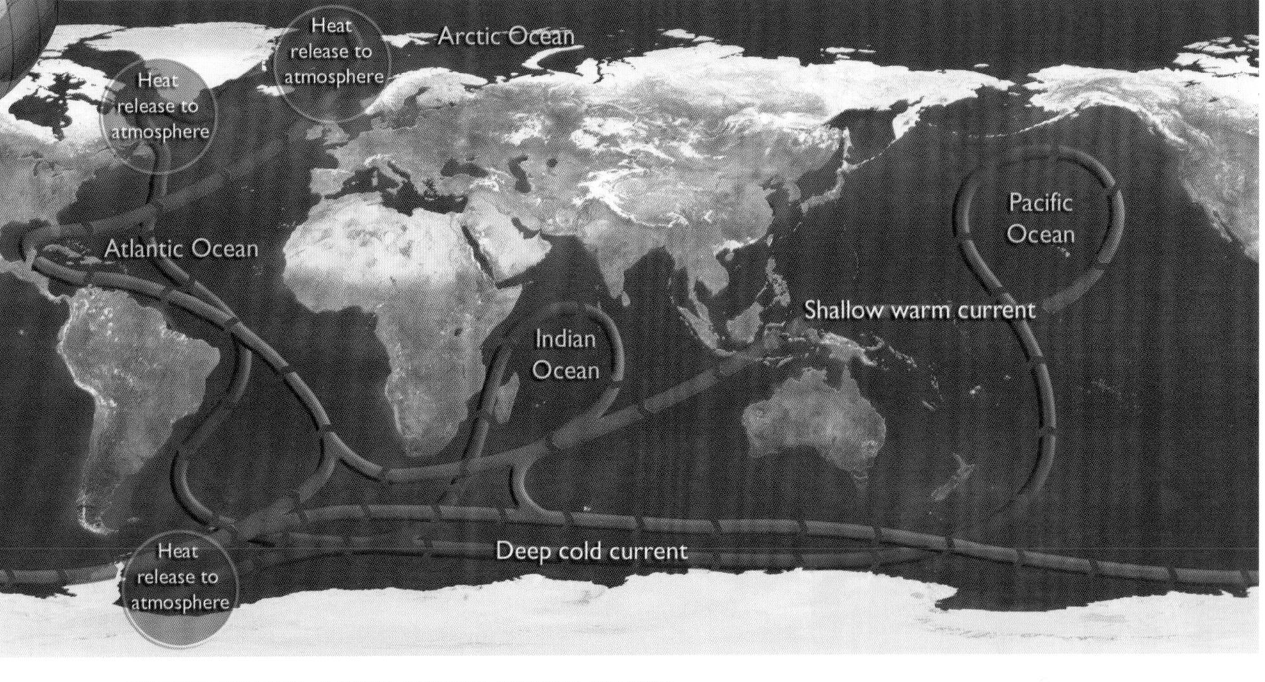

Changes in global ocean circulation can lead to abrupt climate change. Such change can be initiated by increases in arctic precipitation and river runoff, and the melting of arctic snow and ice, because these lead to reduced salinity of ocean waters in the North Atlantic, evidence of which is shown above. Further discussion of this issue is found on pages 36-37.

The Importance of Thresholds

There are many thresholds in the arctic environment, which if crossed, could lead to significant consequences for the region and the world. As human-induced warming continues, the potential exists for various arctic systems to shift to new or unusual states. Such changes might be initiated when a temperature or precipitation threshold is crossed. Records of ancient arctic climate suggest that such changes in some cases occurred abruptly (over a few years) and in other cases more gradually (over several decades or more). Such shifts have the potential to cause the relatively rapid onset of various types of impacts. For example, unusually warm and wet conditions might accelerate outbreaks of pests or infectious diseases.

The onset of the long-term melting of the Greenland Ice Sheet is an example of a threshold that is likely to be crossed during this century. Climate models project that local warming in Greenland will exceed 3˚C during this century. Ice sheet models project that a warming of that magnitude would initiate the long-term melting of the Greenland Ice Sheet. Even if climatic conditions then stabilized, an increase of this magnitude is projected to lead eventually (over centuries) to a virtually complete melting of the Greenland Ice Sheet, resulting in a global sea level rise of about seven meters. The tentative indication from the North Atlantic of an initial slowing of the deep ocean circulation is another example of a possible threshold that might be crossed. If present trends continue, leading to a significant slowdown, the northward oceanic transport of tropical warmth that now moderates European winters could be significantly diminished.

There are also thresholds that can be crossed in the world of living things. For example, nearly half of all white spruce trees at treeline in Alaska show a marked decrease in growth when average July temperature measured at a nearby station exceeds 16˚C. The observed relationship between this temperature threshold and reduced growth suggests that growth of these trees would cease entirely under the amount of warming projected to occur during this century, thus eliminating this population. Similar population crashes in some animal species can result if critical thresholds are passed.

Sudden or unexpected changes pose great challenges for scientists to project as well as for the ability of societies to adapt, and can thus increase vulnerability to significant impacts. While there is still much uncertainty about which of these thresholds will be crossed and exactly when this might occur, records of the past suggest that the possibilities for abrupt changes and new extremes are real.

Rates of Change

The rate at which a change occurs can be as important or more important than the amount of change. When examining impacts of climate change, rates of change must be considered in relation to factors that are important to human society. For example, if thawing of permafrost or increasing coastal erosion were to occur very slowly, people might be able to replace buildings, roads and the like as part of the normal replacement cycle for infrastructure. But if the changes occur rapidly, then the costs of replacement can be high.

The rate of change is also critical to the discussion of abrupt climate change. Evidence from past climates suggests that abrupt changes are most likely to occur when earth's climate is changing rapidly. Since abrupt changes are often most difficult to adapt to, the rate of climate change is clearly a major concern.

Climate models project that local warming in Greenland will exceed 3˚C during this century. Ice sheet models project that a warming of that magnitude would initiate the long-term melting of the Greenland Ice Sheet.

Projected Annual Temperature Change 2070-2090

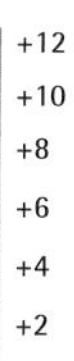

2 Arctic warming and its consequences have worldwide implications.

Importance of the Arctic to Global Climate

The Arctic exerts a special influence over global climate. Integrated over the year, incoming energy from the sun is greatest near the equator and smallest near the poles. Further, because much of the Arctic is covered with snow and ice, a larger fraction of the incoming solar energy is reflected back to space than at lower latitudes, which absorb most of this energy. If not for the atmosphere and oceans moving energy from the tropics to the poles, the tropics would overheat and the polar regions would be much colder than they are. In the Northern Hemisphere, the Atlantic Ocean is the major carrier of the oceanic component of this energy transfer, and as explained below, arctic processes have the potential to have major impacts on the strength of the Atlantic Ocean's circulation.

If not for the atmosphere and oceans moving energy from the tropics to the poles, the tropics would overheat and the polar regions would be much colder than they are.

There are three major mechanisms, or so-called "feedbacks", by which arctic processes can cause additional climate change for the planet. One involves changes in the reflectivity of the surface as snow and ice melt and vegetation cover changes, the second involves changes to ocean circulation as arctic ice melts, adding freshwater to the oceans, and the third involves changes in the amounts of greenhouse gases emitted to the atmosphere from the land as warming progresses.

Feedback 1: Surface Reflectivity

The first feedback involves the snow and ice that cover much of the Arctic. Because they are bright white, snow and ice reflect most of the solar energy that reaches the surface back to space. This is one reason why the Arctic remains so cold. As greenhouse gas concentrations rise and warm the lower atmosphere, snow and ice begin to form later in the autumn and melt earlier in the spring. The melting back of the snow and ice reveals the land and water surfaces beneath, which are much darker, and thus absorb more of the

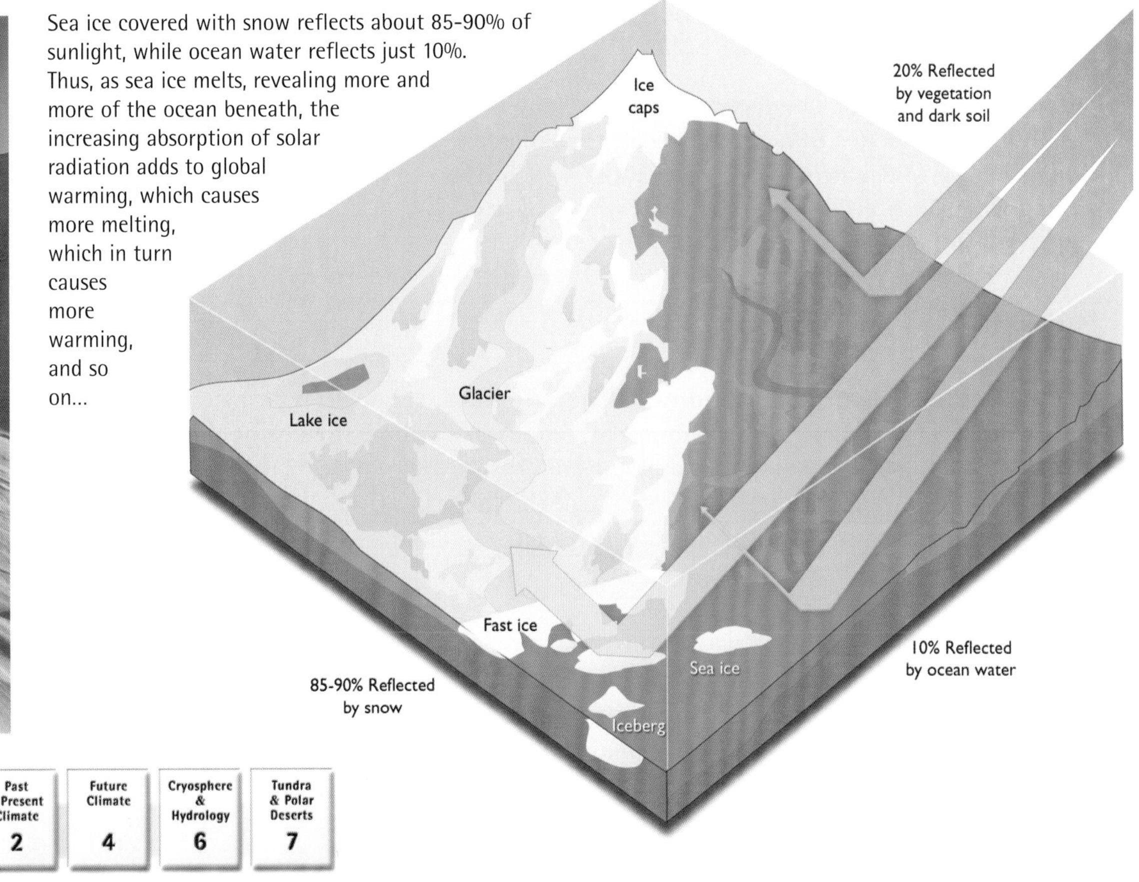

Sea ice covered with snow reflects about 85-90% of sunlight, while ocean water reflects just 10%. Thus, as sea ice melts, revealing more and more of the ocean beneath, the increasing absorption of solar radiation adds to global warming, which causes more melting, which in turn causes more warming, and so on...

sun's energy. This warms the surface further, causing faster melting, which in turn causes more warming, and so on, creating a self-reinforcing cycle by which global warming feeds on itself, amplifying and accelerating the warming trend. This process is already underway in the Arctic with the widespread retreat of glaciers, snow cover, and sea ice. This is one reason why climate change is more rapid in the Arctic than elsewhere. This regional warming also accelerates warming at the global scale.

As melting snow and ice reveal darker land and water surfaces, more of the sun's energy is absorbed, further warming the planet.

Another warming-induced change likely to increase the absorption of solar energy at the earth's surface is that forests are projected to expand northward into areas that are currently tundra. The relatively smooth tundra is much more reflective, especially when covered with snow, than the taller, darker, and more textured forests that are projected to replace it. The reduction in reflectivity that will accompany forest expansion is expected to further increase warming, though more slowly than the snow and ice changes described above. The increased tree growth is expected to absorb more carbon dioxide than the current vegetation, a process that could potentially moderate warming. However, the reduction in reflectivity resulting from forest expansion is likely to exert a larger climatic influence than the carbon uptake effect, thus amplifying warming. Greater vegetation growth also masks snow cover on the ground, further reducing surface reflectivity.

A direct human influence that also decreases reflectivity is that soot is produced when fossil fuels are burned (in addition to the carbon dioxide that is the primary problem). The soot that is carried by the winds and deposited in the Arctic slightly darkens the surface of the otherwise bright white snow and ice, causing them to reflect less of the sun's energy, thus further increasing warming. Soot in the atmosphere also increases solar absorption, further warming the region.

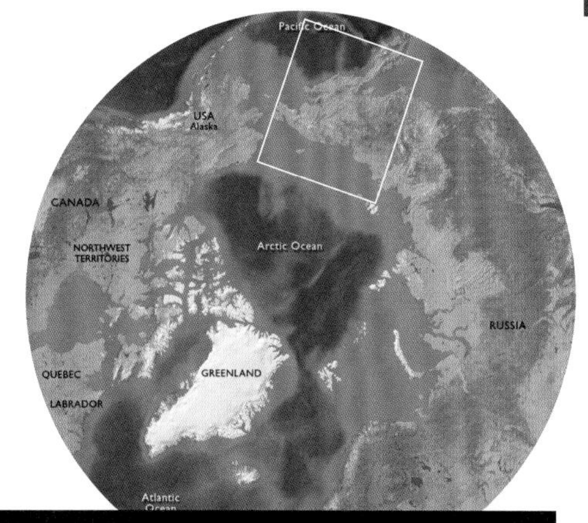

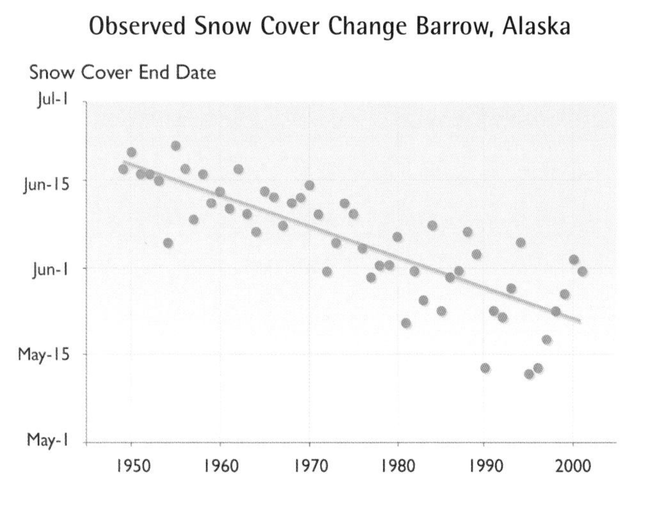

Snow cover extent over arctic land areas has decreased by about 10% over the past 30 years, with the most visible change being an earlier disappearance of snow in spring. One local example is shown in the graph above, for Barrow, Alaska. Over the past 50 years, the snow cover end date has shifted to about one month earlier.

Current Arctic Vegetation

Projected Vegetation, 2090-2100

- Ice
- Polar Desert / Semi-desert
- Tundra
- Boreal Forest
- Temperate Forest
- Grassland

These maps of current and projected vegetation in the Arctic illustrate that forests are projected to overtake tundra and tundra is projected to move into polar deserts. These changes will result in a darker land surface, amplifying warming by absorbing more of the sun's energy and creating a self-reinforcing feedback loop.

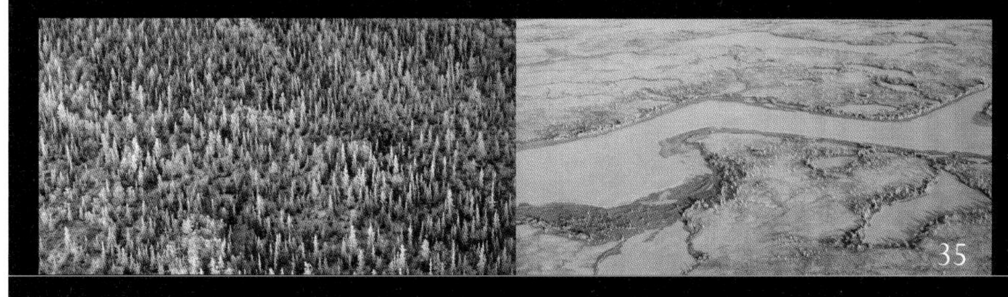

2 Arctic warming and its consequences have worldwide implications.

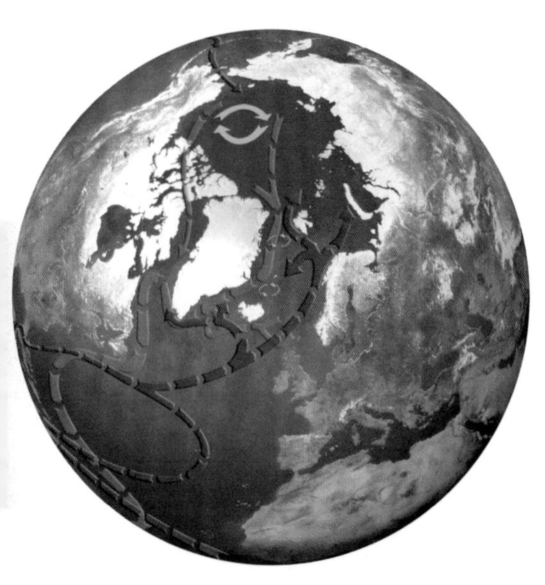

Arctic Thermohaline Circulation

Feedback 2: Ocean Circulation

The second feedback through which arctic processes can amplify changes in global climate is through alterations in ocean circulation patterns. One of the ways the sun's energy is transported from the equator toward the poles is through the globally interconnected movement of ocean waters (see page 32) primarily driven by differences in heat and salt content, known as the thermohaline circulation ("thermo" for heat and "haline" for salt).

At present, the northward extension of the warm Gulf Stream current in the North Atlantic Ocean warms the winds and provides much of the moisture that falls as precipitation over northwestern Europe. As the waters move northward, they become cooler and denser, until they are heavier than the waters below and sink deep into the ocean. This sinking occurs particularly in the seas making up the northern North Atlantic Ocean, and in the Labrador Sea, driving the global thermohaline circulation (sometimes referred to as the "conveyor belt"). This sinking of dense seawater pulls more warm waters northward, helping to provide the heat that keeps Europe warmer in winter than regions of North America that are at the same latitude.

The formation of sea ice also makes the near-surface water saltier and denser as salt is rejected from the ice. In the shallow coastal seas, this water becomes saltier and dense enough to sink. It then flows down the continental shelves into the deep ocean basin, contributing to deep-water formation and the further drawing northward of heat from the tropics. This process is delicately balanced; if the waters are made less salty by an increase in freshwater from runoff and precipitation, or because temperatures are not sufficiently cold to form sea ice, the formation rate of deep water will decrease and less heat from the tropical regions will be pulled northward by the ocean to moderate European winters.

Science Chapters: Past & Present Climate 2 | Future Climate 4 | Cryosphere & Hydrology 6 | Marine Systems 9

Major Arctic Rivers

This map illustrates the major river networks of the Arctic. The thickness of the blue lines represents the relative river discharge, with the thickest lines indicating the rivers with the largest volume. The numbers on the map are in cubic kilometers per year.

Climate change is projected to influence all of these processes, causing more freshwater to be carried by rivers to the Arctic Ocean due to melting of glaciers and increases in precipitation, and warming the ocean and reducing the rate at which sea-ice formation creates saltier and denser water. As high-latitude waters thus become less salty, they float on top of the saltier waters below, capping them in much the same way as a layer of oil rests above a layer of water. This inhibits vertical mixing, slows the formation of deepwater, and slows the thermohaline circulation.

Slowing the thermohaline circulation would have several important global effects. Because the oceanic overturning is an important mechanism for carrying carbon dioxide to the deep ocean, slowing of this circulation would allow the carbon dioxide concentration in the atmosphere to build up more rapidly, leading to more intense and longer-lasting global warming. Slowing of this ocean circulation would also slow the northward transport of heat by Atlantic Ocean currents, thereby slowing the rate of warming in the region, and perhaps causing regional cooling for several decades, even as the rest of the planet warms more rapidly.

Reducing bottom water formation in the Arctic would also reduce the amount of heat and nutrients carried back toward the surface elsewhere in the world by the upward moving components of the thermohaline circulation. This would increase the rate of sea-level rise (due to greater thermal expansion) while reducing the supply of nutrients available to near-surface marine life and the transport of carbon to the deep ocean as carbon-containing living things die and sink. Thus, what happens in the Arctic will have ramifications around the world.

Rising River Flows

The Arctic Ocean is the most river-influenced ocean in the world, receiving 11% of the world's river flow yet containing only 1% of the global volume of seawater. An overall increase has been observed in freshwater flow from rivers to the Arctic Ocean over the past 100 years, with the largest increases taking place in winter and since 1987, in accordance with the largest increases in air temperature. Springtime peak flows are occurring earlier on many rivers. For the next 100 years, models project 10-25% increases in annual river discharge, with greater increases in winter and spring. If summer warming increases losses due to evaporation, it is possible that river levels and flow rates will decrease from present values during summer.

Rising Winter River Discharge

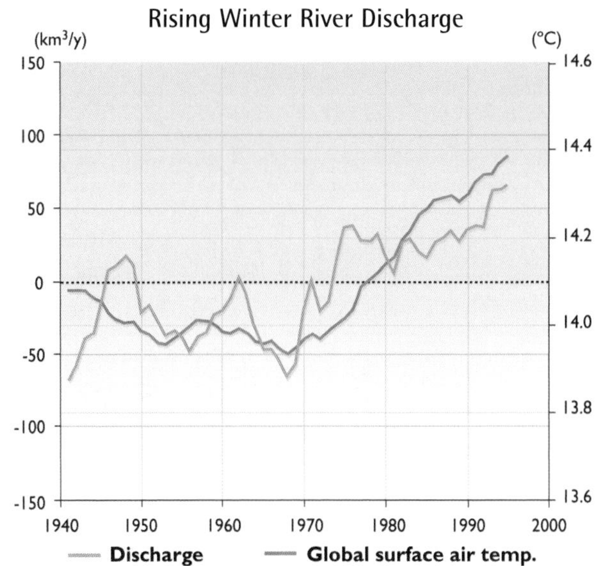

The purple line shows departures from the long-term average of European river discharge in winter (December through March), and the blue line shows changes in global average surface air temperature.

Key Finding #2

2 Arctic warming and its consequences have worldwide implications.

Feedback 3: Greenhouse Gas Emissions

A third feedback through which arctic processes can affect global climate change is by modifying the exchange of greenhouse gases between the atmosphere and arctic soils and sediments, which are likely to be affected as air and water warm.

Methane and Carbon Dioxide from Permafrost

Carbon is currently trapped as organic matter in the permafrost (frozen soil) that underlies much of the Arctic. Large amounts of carbon accumulate particularly in the vast waterlogged peat bogs of Siberia and parts of North America. During the summer, when the surface layer of the permafrost thaws, organic matter in this layer decomposes, releasing methane and carbon dioxide to the atmosphere. Warming increases these releases, and can create an amplifying feedback loop whereby more warming causes additional releases, which would cause more warming, and so on. The potential magnitude of these releases is affected by soil moisture and numerous other factors, and is thus subject to substantial uncertainties.

Permafrost Distribution

Continuous
Discontinuous
Sporadic
Subsea

Subsea permafrost in the Arctic occurs in the wide continental shelf area. Narrow zones of coastal permafrost are probably present along most arctic coasts.

Methane and Carbon Dioxide in Forests and Tundra

The boreal forests and arctic tundra contain some of the world's largest land-based stores of carbon, primarily in the form of plant material in the forests and as soil carbon in the tundra. Methane is about 23 times as potent at trapping heat in the earth's atmosphere as carbon dioxide (by weight, over a 100-year time horizon). Methane is produced by the decomposition of dead plant material in wet soils such as mires and tundra ponds. The release of methane to the atmosphere is generally accelerated by rising temperatures and precipitation, although in areas where drying occurs, methane may be absorbed by forest and tundra soils. Carbon dioxide is released by decomposition in soils in drier areas and by burning of trees in fires. Increasing temperatures will lead to faster decomposition initially, but the likely replacement of arctic vegetation by more productive vegetation from the south would be expected to result in a greater uptake of carbon, except in disturbed and particularly dry areas. It is not known whether the net effect of these changes will be a greater overall carbon uptake as climate change proceeds, though recent studies suggest that over the Arctic as a whole, more productive vegetation will probably increase carbon storage in ecosystems.

Methane Hydrates in the Coastal Arctic Ocean

Vast amounts of methane, in a solid icy form called methane hydrates or clathrates, are trapped in permafrost and at shallow depths in cold ocean sediments. If the temperature of the permafrost or water at the seabed rises a few degrees, it could initiate the decomposition of these hydrates, releasing methane to the atmosphere. The release of methane from this source

Subsea Methane Hydrates

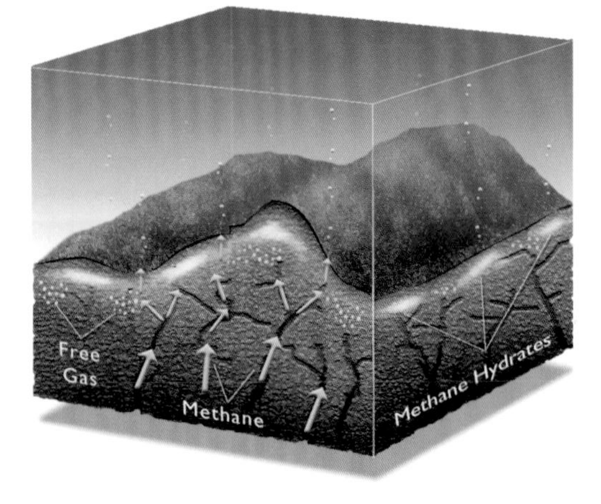

is a less certain outcome of climate change than the other emissions discussed here because it would probably require greater warming and take longer to occur. If such releases did occur, however, the climate impacts could be very large.

Carbon Uptake in the Oceans

To date, the Arctic Ocean has not played a very large part in the global carbon budget because absorption of carbon dioxide from the air has been limited by the sizeable ice cover and because uptake to support biological productivity under perennial sea ice has not been very significant compared to elsewhere in the world's oceans. Under warmer climate conditions, however, it is possible that the amount of carbon taken up by the Arctic Ocean could increase significantly. With less sea ice, more carbon dioxide is likely to be absorbed by the very cold waters, and, as dense water is formed during the seasonal formation of sea ice, the additional carbon dioxide could be carried downward. In addition, increased biological productivity in the open waters could lead to more carbon being carried down as living things die and sink, especially if increased runoff adds to the amount of available nutrients. While these changes are likely to be important regionally, the total area is not large enough to significantly reduce global atmospheric carbon dioxide concentrations.

This schematic illustrates changes in the cycling of carbon in the Arctic as climate warms. For example, beginning at the left of the figure, the boreal forest absorbs CO_2 from the atmosphere and this is expected to increase, although forest fires and insect damage will increase in some areas, releasing more carbon to the atmosphere. Increasing amounts of carbon will also move from the tundra to ponds, lakes, rivers, and the continental shelves in the form of carbon dissolved in water (dissolved organic carbon (DOC), dissolved inorganic carbon (DIC), and particulate organic carbon (POC)).

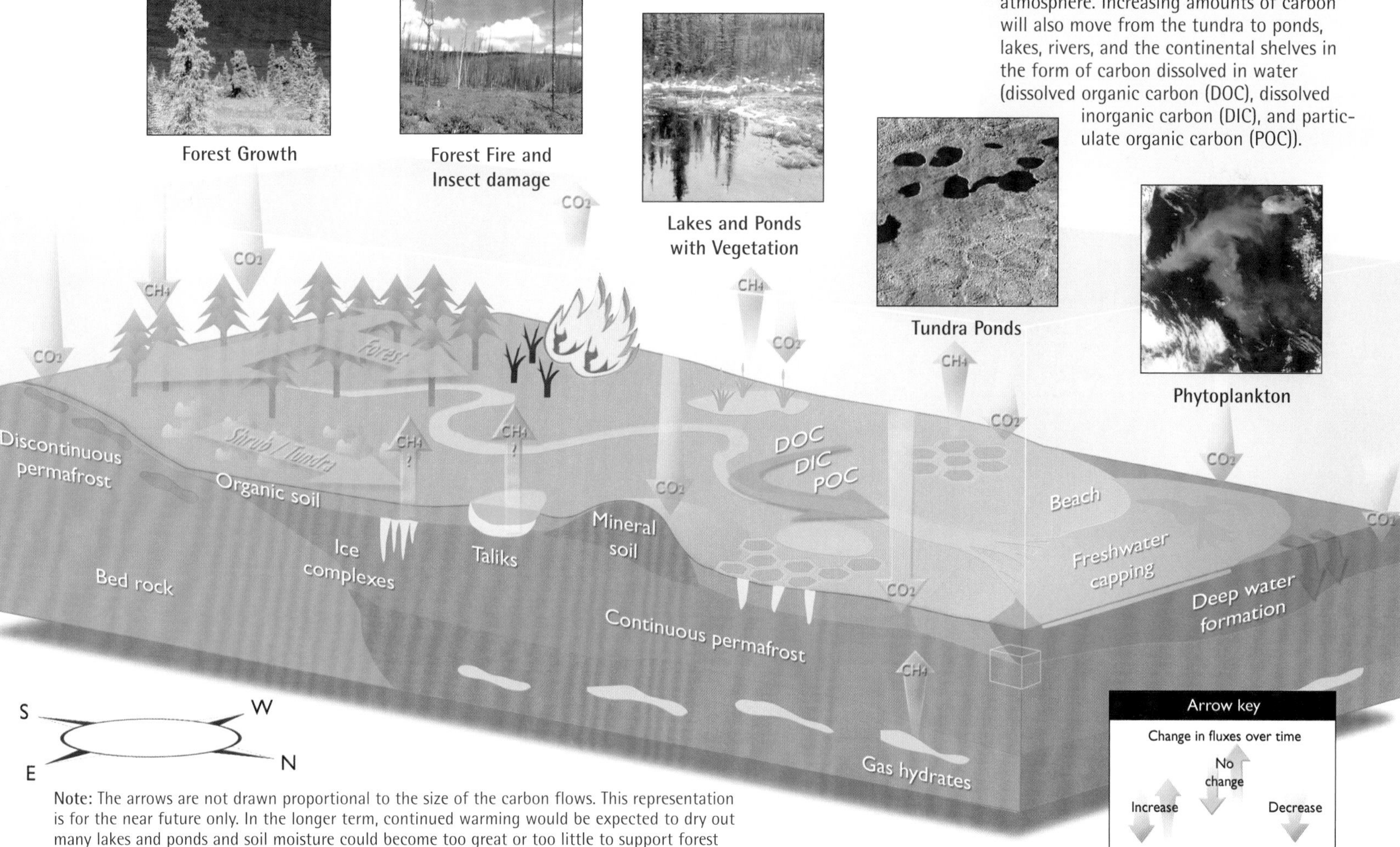

Note: The arrows are not drawn proportional to the size of the carbon flows. This representation is for the near future only. In the longer term, continued warming would be expected to dry out many lakes and ponds and soil moisture could become too great or too little to support forest expansion, resulting in large releases of carbon to the atmosphere.

2 Arctic warming and its consequences have worldwide implications.

Melting Glaciers Contribute to Global Sea-level Rise

The total volume of land-based ice in the Arctic has been estimated to be about 3 100 000 cubic kilometers, which corresponds to a sea-level equivalent of about eight meters. Most arctic glaciers and ice caps have been in decline since the early 1960s, with this trend speeding up in the 1990s. A small number of glaciers, especially in Scandinavia, have gained mass as increased precipitation outpaced the increase in melting in a few areas.

The Greenland Ice Sheet dominates land ice in the Arctic. Maximum surface-melt area on the ice sheet increased on average by 16% from 1979 to 2002, an area roughly the size of Sweden, with considerable variation from year to year. The total area of surface melt on the Greenland Ice Sheet broke all records in 2002, with extreme melting reaching up to a record 2000 meters in elevation. Satellite data show an increasing trend in melt extent since 1979. This trend was interrupted in 1992, following the eruption of Mt. Pinatubo, which created a short-term global cooling as particles spewed from the volcano reduced the amount of sunlight that reached the earth.

Recent studies of glaciers in Alaska indicate an accelerated rate of melting. The associated sea-level rise is nearly double the estimated contribution from the Greenland Ice Sheet during the past 15 years. This rapid retreat of Alaska's glaciers represents about

"That year [2002] the melt was so early and so intense – it really jumped out at me. I'd never seen the seasonal melt occur that high on the ice sheet before, and it had never started so early in the spring."

Konrad Steffan
University of Colorado, USA

Greenland Ice Sheet Melt Extent

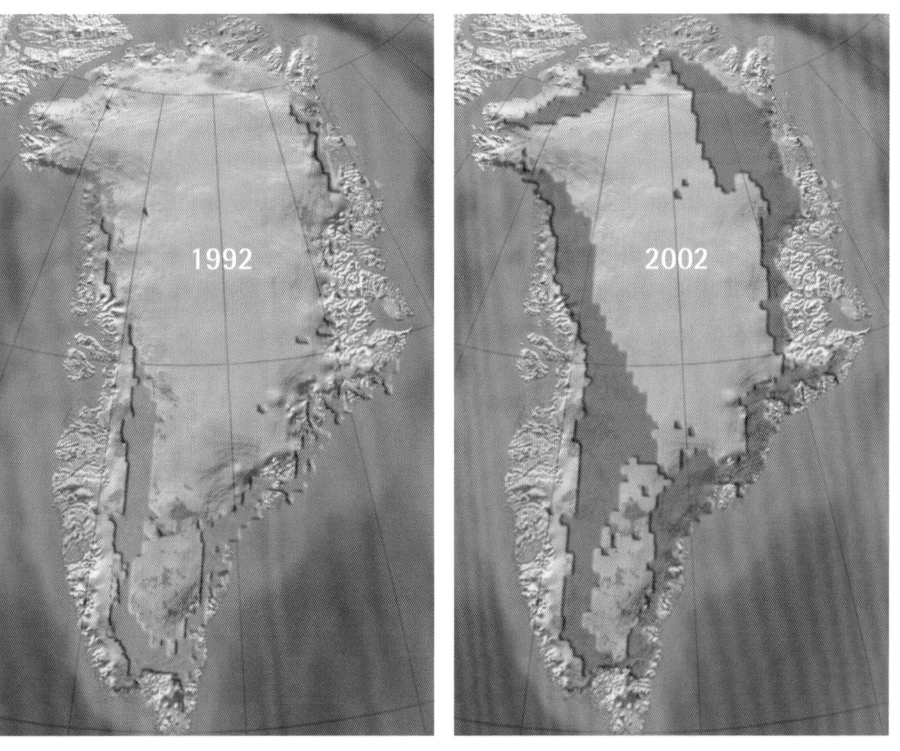

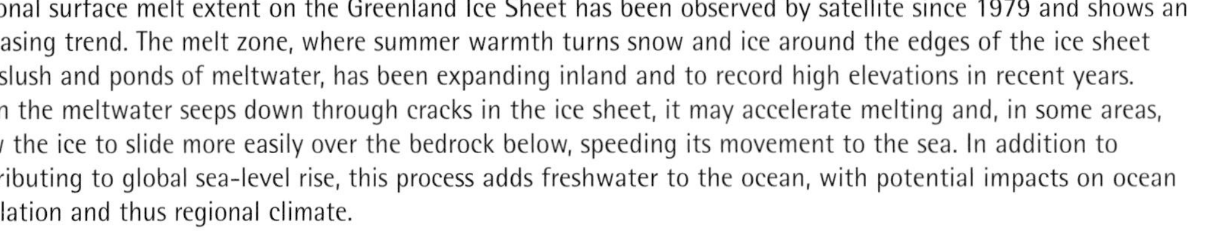

Seasonal surface melt extent on the Greenland Ice Sheet has been observed by satellite since 1979 and shows an increasing trend. The melt zone, where summer warmth turns snow and ice around the edges of the ice sheet into slush and ponds of meltwater, has been expanding inland and to record high elevations in recent years. When the meltwater seeps down through cracks in the ice sheet, it may accelerate melting and, in some areas, allow the ice to slide more easily over the bedrock below, speeding its movement to the sea. In addition to contributing to global sea-level rise, this process adds freshwater to the ocean, with potential impacts on ocean circulation and thus regional climate.

Greenland Ice Sheet Melt Extent
(Maximum melt extent 1979 - 2002)

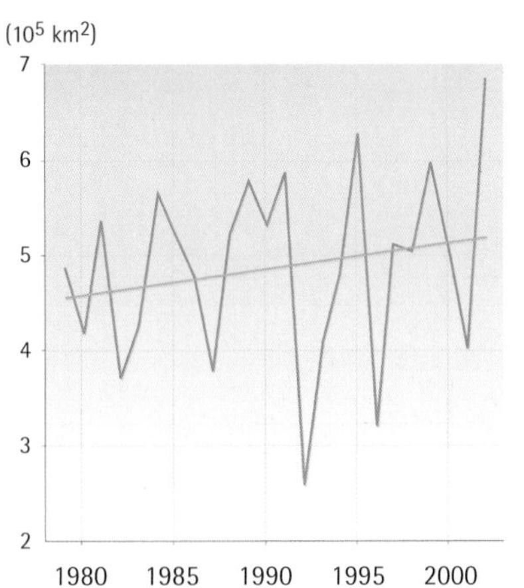

half of the estimated loss of mass by glaciers worldwide, and the largest contribution by glacial melt to rising sea level yet measured.

Projections from global climate models suggest that the contribution of arctic glaciers to global sea-level rise will accelerate over the next 100 years, amounting to roughly four to six centimeters by 2100. Recent research suggests that this estimate should be higher due to the increase in arctic glacial melt during the past two decades.

Over the longer term, the arctic contribution to global sea-level rise is projected to be much greater as ice sheets continue to respond to climate change and to contribute to sea-level rise for thousands of years. Climate models indicate that the local warming over Greenland is likely to be one to three times the global average. Ice sheet models project that local warming of that magnitude would eventually lead to a virtually complete melting of the Greenland Ice Sheet, with a resulting sea-level rise of about seven meters.

"Glaciers are very notably receding and the place names are no longer consistent with the appearance of the land. For example, Sermiarsuussuaq ('the Smaller Large Glacier'), which previously stretched out to the sea, no longer exists."

Uusaqqak Qujaukitsoq
Qaanaaq, Greenland

Key Finding #2

McCall Glacier Retreat
Brooks Range, Alaska

1958

2003

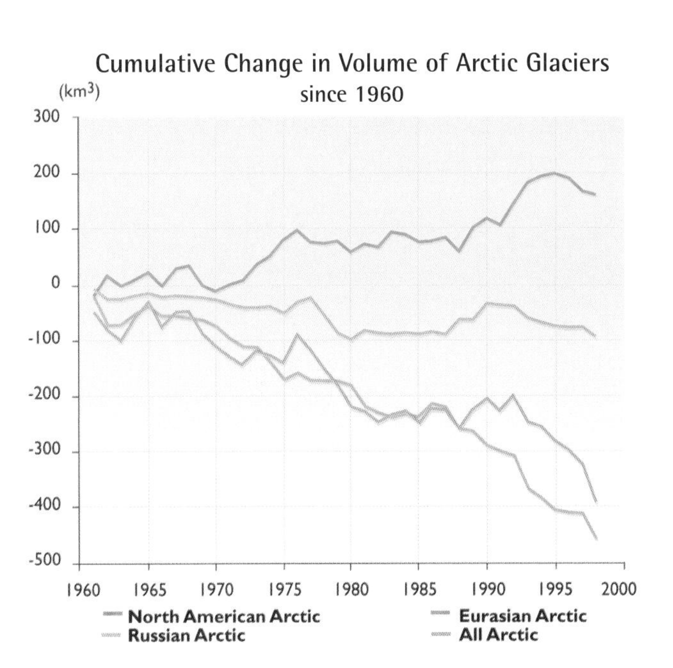

For the Arctic as a whole, there was a substantial loss in glacial volume from 1961 to 1998. Glaciers in the North American Arctic lost the most mass (about 450 km^3), with increased loss since the late 1980s. Glaciers in the Russian Arctic have also had large losses (about 100 km^3). Glaciers in the European Arctic show an increase in volume because increased precipitation in Scandinavia and Iceland added more to glacial mass than melting removed over that period.

2 Arctic warming and its consequences have worldwide implications.

Sea-level rise is projected to have serious implications for coastal communities and industries, islands, river deltas, harbors, and the large fraction of humanity living in coastal areas worldwide.

Impacts of Global Sea-level Rise

Sea-level rise has the potential for significant impacts on societies and ecosystems around the world. Climate change causes sea level to rise by affecting both the density and the amount of water in the oceans. First and most significantly, water expands as it warms, and less-dense water takes up more space. This "thermal expansion" is projected to be the largest component of sea-level rise over the next 100 years and will persist for many centuries. Secondly, warming increases melting of glaciers and ice caps (land-based ice), adding to the amount of water flowing into the oceans.

Global average sea level rose almost three millimeters per year during the 1990s, up from about two millimeters per year in the several decades prior to that. This rate is, in turn, 10 to 20 times faster than the estimated rate of rise over the past few thousand years. The primary factors contributing to this rise are thermal expansion due to ocean warming and melting of land-based ice that increases the total amount of water in the ocean.

Global average sea level is projected to rise 10 to 90 centimeters during this century, with the rate of rise accelerating as the century progresses. Over the longer term, much larger increases in sea level are projected. Sea-level rise is expected to vary around the globe, with the largest increases projected to occur in the Arctic, in part due to the projected increase in freshwater input to the Arctic Ocean and the resulting decrease in salinity and thus density.

Sea-level rise is projected to have serious implications for coastal communities and industries, islands, river deltas, harbors, and the large fraction of humanity living in coastal areas worldwide. Sea-level rise will increase the salinity of bays and estuaries. It will increase coastal erosion, especially where coastal lands are soft rather than rocky.

Observed Global Sea Level Rise

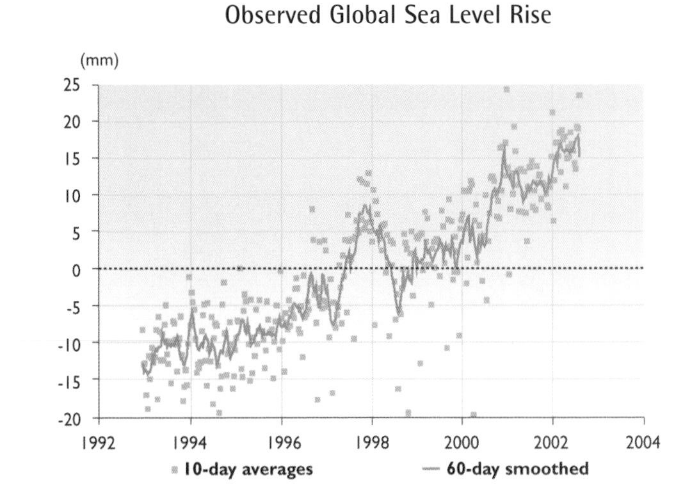

These data, from a satellite launched in 1992, show the rise in global average sea level over the past decade.

Projected Global Sea Level Rise

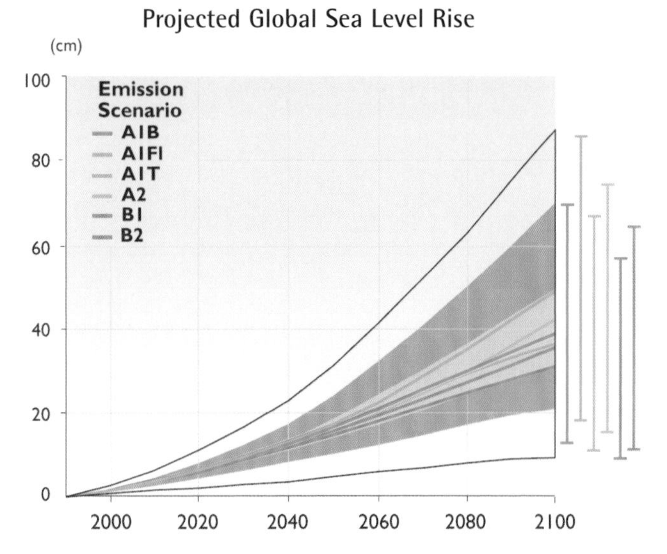

The graph shows future increases in global average sea level in meters as projected by a suite of climate models using six IPCC emissions scenarios. The bars at right show the range projected by a group of models for the designated emissions scenarios.

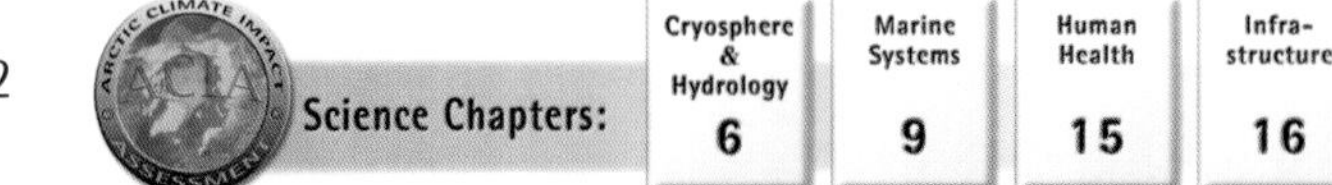

Extensive coastal lowlands and delta areas contain important ecosystems that will be affected by rising sea levels. Wetlands will be forced farther inland, and the incidence of coastal floods will increase.

The impacts of sea-level rise are likely to be most severe along gently sloping coastal lands, inland areas bordering estuaries, and coastlines that are subsiding due to tectonic forces, sedimentation, or extraction of oil or groundwater. Low-lying islands in the Pacific Ocean (Marshall, Kiribati, Tuvalu, Tonga, Line, Micronesia, Cook), Atlantic Ocean (Antigua, Nevis), and Indian Ocean (Maldives), are very likely to be severely affected.

In Bangladesh, about 17 million people live less than one meter above sea level and are already vulnerable to flooding. In Southeast Asia, a number of very large cities including Bangkok, Bombay, Calcutta, Dhaka, and Manila (each with populations greater than five million), are located on coastal lowlands or on river deltas. In the United States, Florida and Louisiana are particularly susceptible to impacts of future sea-level rise.

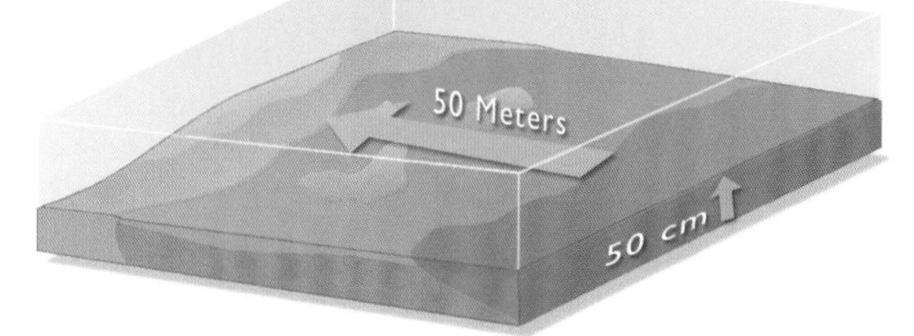

A 50-cm rise in sea level will typically cause a shoreward retreat of coastline of 50 meters if the land is relatively flat (like most coastal plains), causing substantial economic, social, and environmental impacts.

Projected Contribution of Arctic Land Ice to Sea-level Change

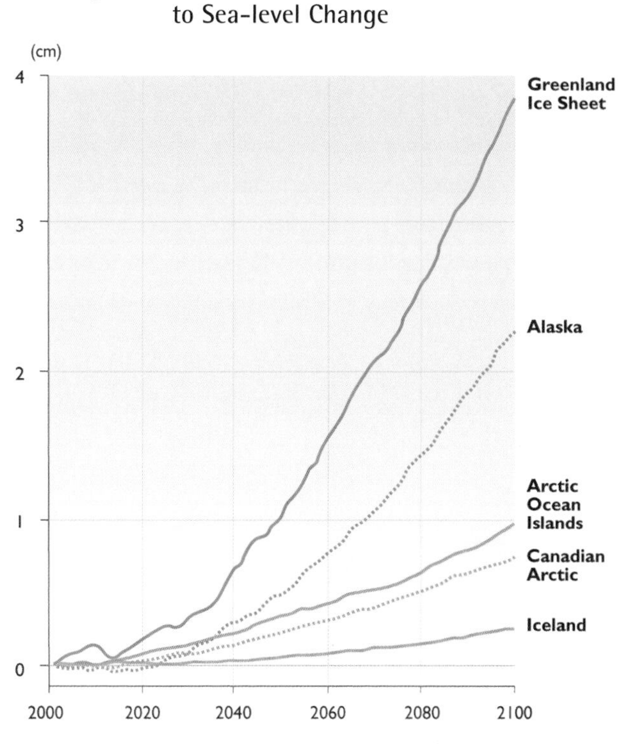

This chart compares the projected contributions to sea-level change due to melting of land-based ice in various parts of the Arctic. The Greenland Ice Sheet is projected to make the largest contribution because of its size. Although Alaska's glaciers cover a much smaller area, they are also projected to make a large contribution. The total contribution of melting land-based ice in the Arctic to global sea-level rise is projected to be about 10 cm by 2100. The primary driver of sea-level rise is thermal expansion due to ocean warming, and that is not included in this chart.

Areas in Florida Subject to Inundation with 100 Centimeter Sea Level Rise

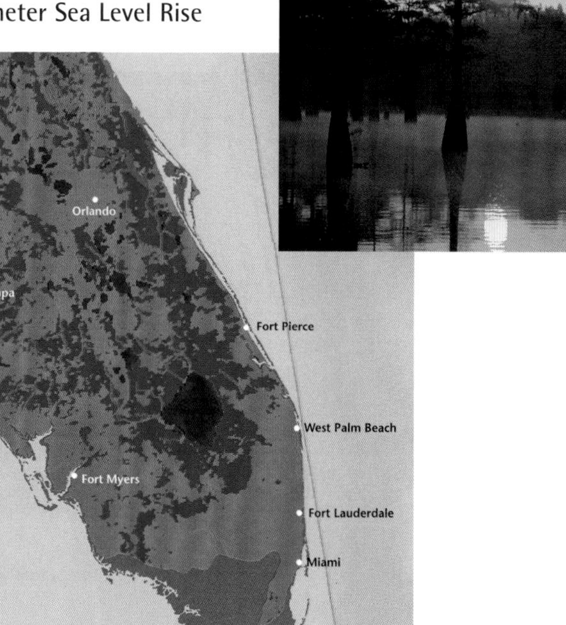

2 Arctic warming and its consequences have worldwide implications.

Access to Arctic Resources Will Change

The Arctic provides natural resources to the world, and climate change will affect these resources in a variety of ways that are examined in more detail throughout this report. Arctic resources have economic value in trade; whales, seals, birds, and fish have long been sold in more southerly markets. Arctic seas contain some of the world's oldest and most productive commercial fishing grounds, which provide significant harvests for many arctic countries, as well as for the rest of the world. For example, Norway is one of the world's largest exporters of fish.

The Arctic has significant oil and gas reserves, most of them located in Russia, with additional fields in Canada, Alaska, Greenland, and Norway. The Arctic also holds large stores of minerals, ranging from gemstones to fertilizers. Russia extracts the largest quantities of minerals, with Canada and Alaska also having significant extraction industries, providing raw materials to the world's economy. Marine access to oil, gas, and minerals is likely to be enhanced in many places in a warmer Arctic, with positive impacts for some and negative impacts for others. Access to resources by land is likely to be hampered in many places due to a shortening of the season during which the ground is sufficiently frozen for travel.

Marine access to oil, gas, and minerals is likely to be enhanced in many places in a warmer Arctic.

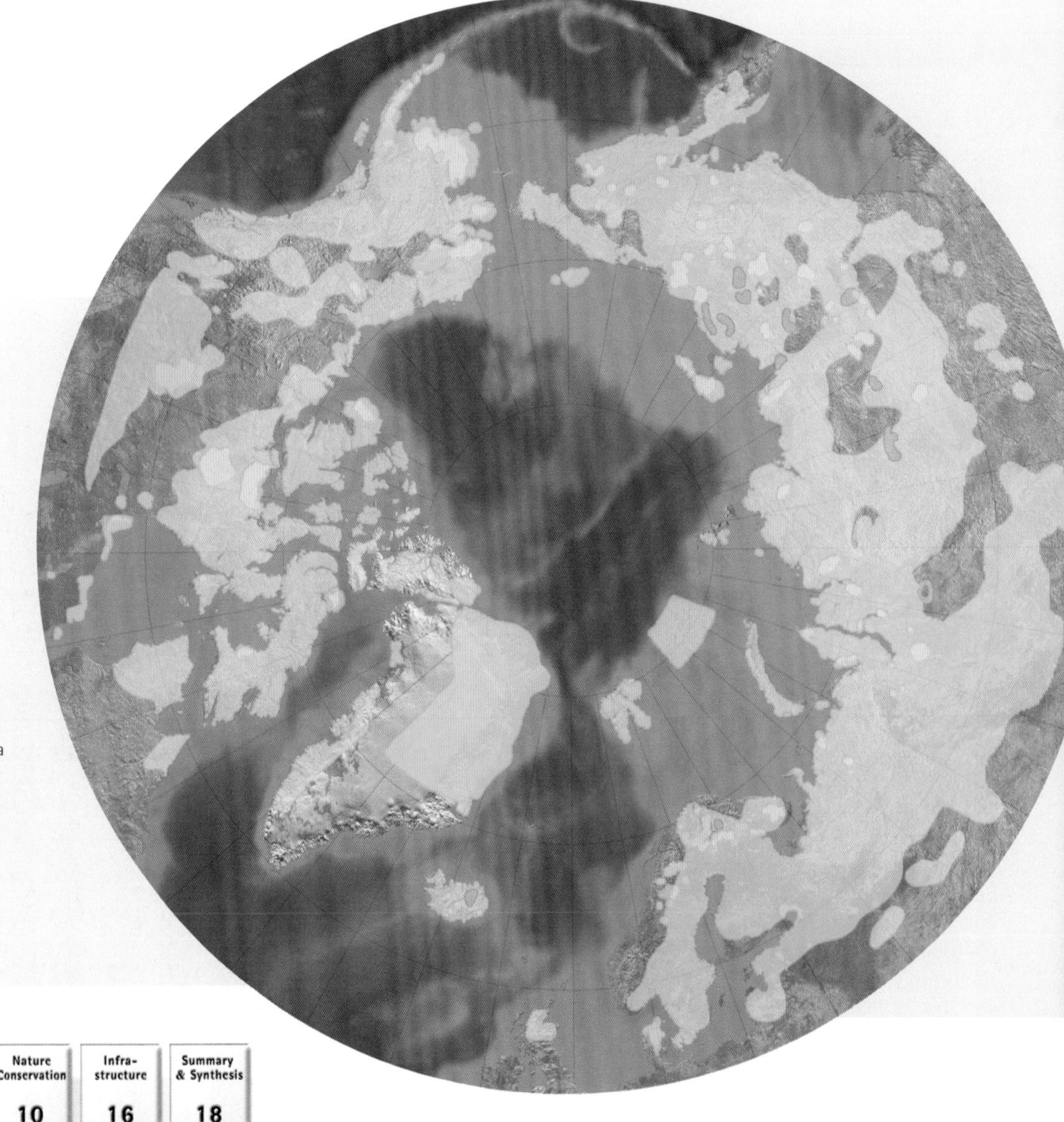

Arctic Protected Areas

Strategies for conserving arctic biodiversity by establishing protected areas are important for defending natural habitats against direct human development, but they do not protect against a changing climate. This map indicates how climate change will affect currently protected areas, putting at risk the living resources these areas were designed to protect.

- Strict Nature Reserve / Wilderness Area National Park
- Natural Monument Habitat / Species Management Area
- Protected Landscape / Seascape Managed Resource Protected Area
- Areas with Predicted Future Changes in Vegetation

Arctic Ecosystem Changes Will Reverberate Globally

Climate-related changes in arctic ecosystems will not just have consequences for local people and other living things that depend on these systems for food, habitat, and other goods and services, but will have impacts at the global level because of the many linkages between the Arctic and regions further south. Many species from around the world depend on summer breeding and feeding grounds in the Arctic, and climate change will alter some of these habitats significantly.

For example, several hundred million birds migrate to the Arctic each summer and their success in the Arctic determines their populations at lower latitudes. Important breeding and nesting areas are projected to decrease sharply as treeline advances northward, encroaching on tundra, and because the timing of bird arrival in the Arctic might no longer coincide with the availability of their insect food sources. At the same time, sea-level rise will erode tundra extent from the north in many areas, further shrinking important habitat for many living things. A number of bird species, including several globally endangered seabird species, are projected to lose more than 50% of their breeding area during this century.

Many species from around the world depend on summer breeding and feeding grounds in the Arctic, and climate change will alter some of these habitats significantly.

Migratory Bird Flyways

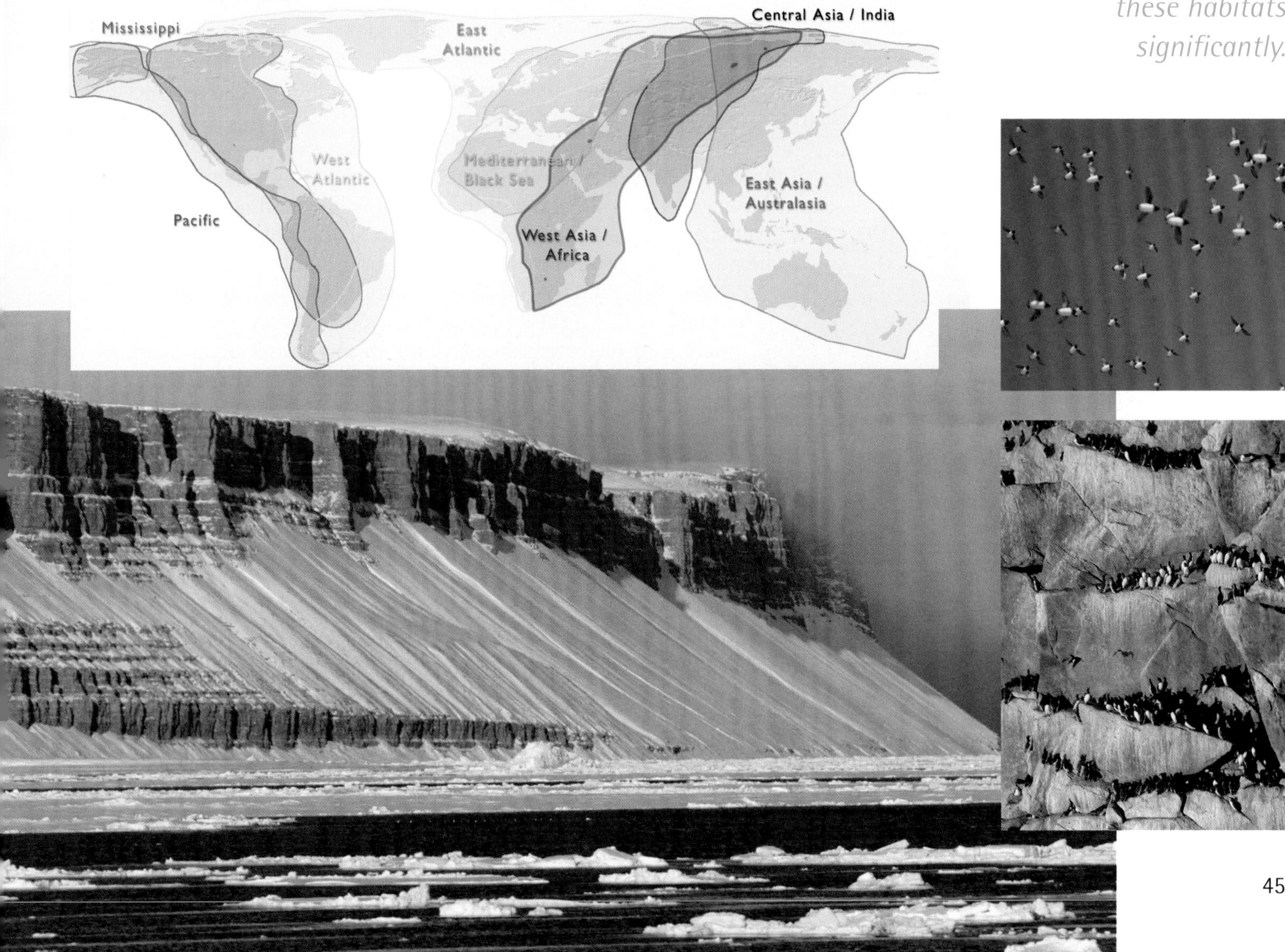

3 Arctic vegetation zones are very likely to shift, causing wide-ranging impacts.

Shifting Vegetation Zones

Climate-induced changes in arctic landscapes are important to local people and animals in terms of food, fuel, culture, and habitat. These changes also have the potential for global impacts because many processes related to arctic landscapes affect global climate and resources. Some changes in arctic landscapes are already underway and future changes are projected to be considerably greater.

The major arctic vegetation zones include the polar deserts, tundra, and the northern part of the boreal forest. The northern-most zone that covers most of the high arctic is the polar desert, characterized by open patches of bare ground and an absence of even the smallest woody shrubs. Although polar desert vegetation is quite sparse, musk ox and small sub-species of caribou/reindeer are found in this zone. Tundra is characterized by low shrub vegetation.

Climate change is projected to cause vegetation shifts because rising temperatures favor taller, denser vegetation.

Climate change is projected to cause vegetation shifts because rising temperatures favor taller, denser vegetation, and will thus promote the expansion of forests into the arctic tundra, and tundra into the polar deserts. The timeframe of these shifts will vary around the Arctic. Where suitable soils and other conditions exist, changes are likely to be apparent in this century. Where they do not, the changes can be expected to take longer. These vegetation changes, along with rising sea levels, are projected to shrink tundra area to its lowest extent in at least the past 21 000 years, greatly reducing the breeding area for many birds and the grazing areas for land animals that depend on the open landscape of tundra and polar desert habitats. Not only are some threatened species very likely to become extinct, some currently widespread species are projected to decline sharply.

Northward Shifting Treeline

Projected Treeline

Present Treeline

Many of the adaptations that enable plants and animals to survive in this cold environment also limit their ability to compete with invading species responding to climate warming. The rapid rate of projected changes in climate also presents special difficulties to the ability of many species to adapt. The primary response of arctic plants and animals to warming is thus expected to be relocation. As species from the south shift their ranges northward, in some cases by as much as 1000 kilometers, they are very likely to displace some arctic species (whose northward shifts are hindered by the Arctic Ocean). Such range displacements have already begun among some bird, fish, and butterfly species. Seabirds, mosses, and lichens are among the groups expected to decline as warming increases. The Arctic is an important global storehouse of diversity for mosses and lichens, containing 600 species of mosses and 2000 species of lichens, more than anywhere else on earth.

The total number of species in the Arctic is projected to increase under a warmer climate due to the influx of species from the south. But entire communities and ecosystems do not shift intact. Rather, the range of each species shifts in response to its sensitivity to climate change, its mobility, its lifespan, and the availability of appropriate soil, moisture, and other needs. The ranges of animals can generally shift much faster than those of plants, and large migratory animals such as caribou can move much more readily than small animals such as lemmings. In addition, migratory pathways must be available, such as northward flowing rivers as conduits for fish species from the southern part of the region. Some migratory pathways may be blocked by development. All of these variations result in the break-up of current communities and ecosystems and the formation of new communities and ecosystems, with unknown consequences.

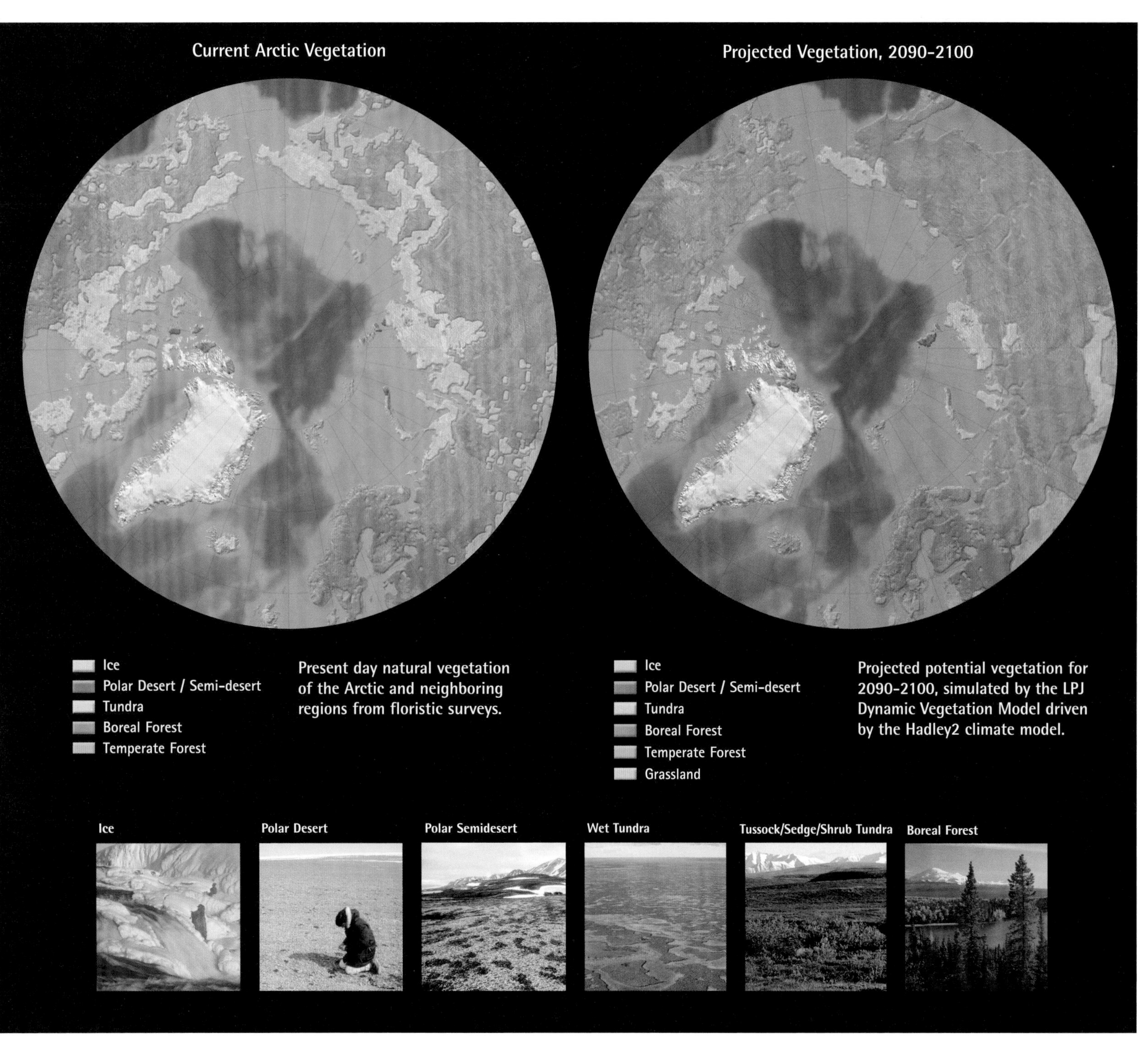

Present day natural vegetation of the Arctic and neighboring regions from floristic surveys.

Projected potential vegetation for 2090-2100, simulated by the LPJ Dynamic Vegetation Model driven by the Hadley2 climate model.

3 Arctic vegetation zones are very likely to shift, causing wide-ranging impacts.

Studies suggest that the increase in absorption of solar radiation will dominate over the increase in carbon storage, resulting in a net increase in warming.

Countervailing Forces on Climate

The projected reduction in tundra and expansion of forest will cause a decrease in surface reflectivity, amplifying global warming because the newly forested areas are darker and more textured and thus will absorb more solar radiation than the lighter, smoother tundra. For example, black spruce is the least reflective of any vegetation type and is likely to be a large part of the mix of new trees in North America. In addition, expanding forests will mask highly reflective snow. The darkening of the surface that results from these changes will create a feedback loop whereby more warming will lead to more tree establishment and forest cover, which will cause more warming, and so on.

On the other hand, the expanding forest vegetation will be more biologically productive than the existing tundra vegetation, and the tundra will be more productive than the polar deserts it displaces. Model results suggest that this could increase carbon storage, slightly moderating the projected amount of warming. The net effect of these countervailing forces involves multiple competing influences that are not fully understood. However, recent studies suggest that the increase in absorption of solar radiation will dominate over the increase in carbon storage, resulting in a net increase in warming.

Desertification: A Potential "Surprise"

Because changes in climate alter many variables and their interrelationships, it is often difficult to project all of the interactive effects on the environment, especially over the long term. While there is high confidence that temperatures will rise and total annual precipitation will increase, it is not known whether the increase in precipitation will

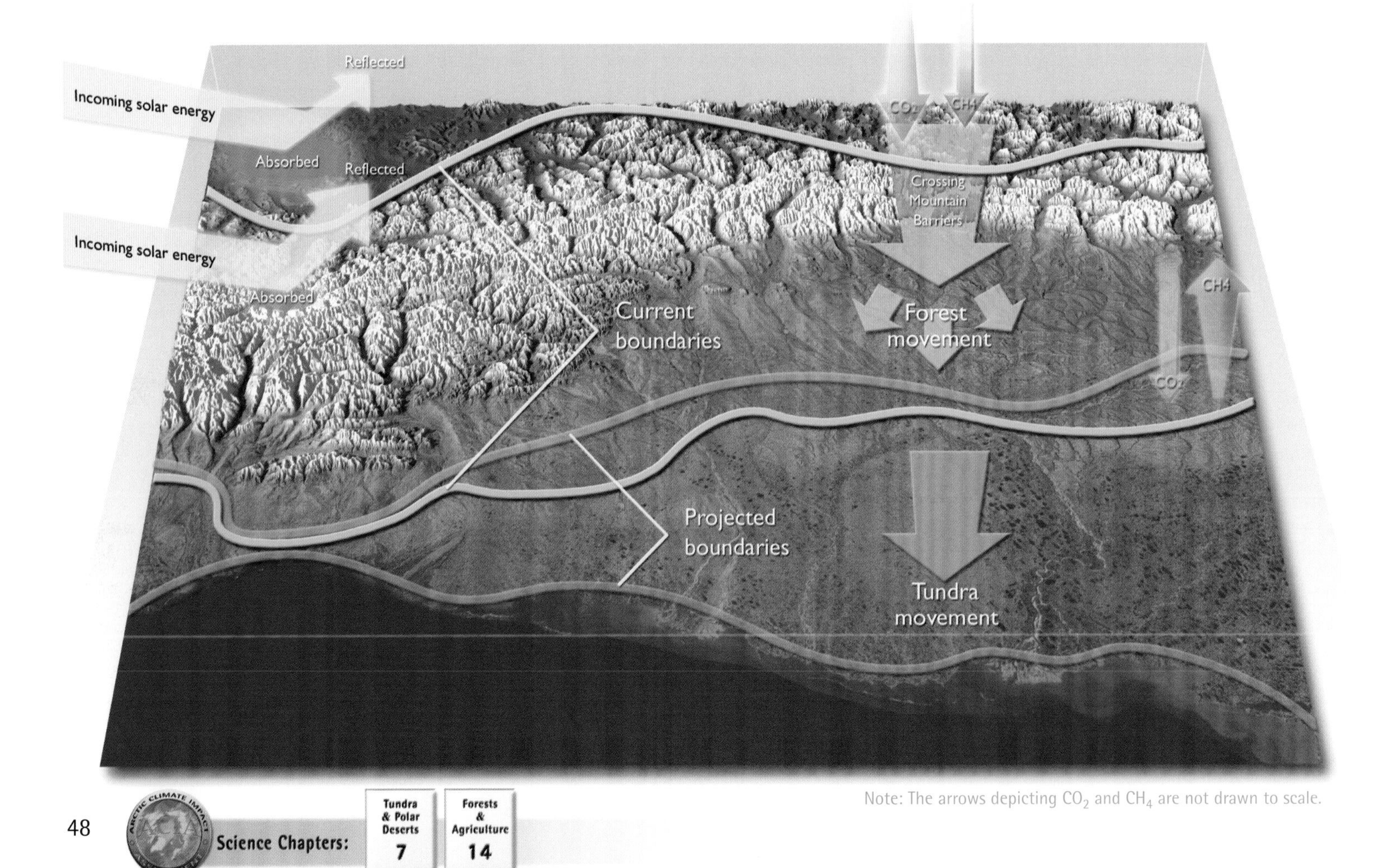

Note: The arrows depicting CO_2 and CH_4 are not drawn to scale.

keep up with the warming in all areas and for all seasons. Because the rate of evaporation increases as temperatures climb, if precipitation does not rise enough to keep up with that increase, land areas would dry out.

Another complexity is the degree of permafrost thawing and the subsequent drainage of water from the land. For example, summer thawing of the active layer of permafrost (the top layer that thaws in summer and freezes in winter) in Barrow, Alaska is now resulting in a great deal of water at the surface. However, this moisture could be lost if the depth of the active layer increases as projected. This is very likely to happen in parts of the Arctic; areas that were not glaciated 10 000 years ago and have fine-grained wind-deposited soils on top of permafrost are particularly prone to drying and erosion. Records of past climatic conditions suggest that this mechanism occurred in the cold and dry tundra-steppe areas of Siberia and Alaska. These processes are likely to lead to an initial greening, followed by desertification in some areas as warming continues.

Polar desert

This series, from a site in northern Sweden similar to that in the photograph, shows that the layer of ground that thaws each summer is growing progressively deeper in recent years as climate has warmed. The red areas are those in which the ground thaws to a depth of 1.1 meters or more. The time series illustrates the rapid disappearance of discontinuous permafrost in this area.

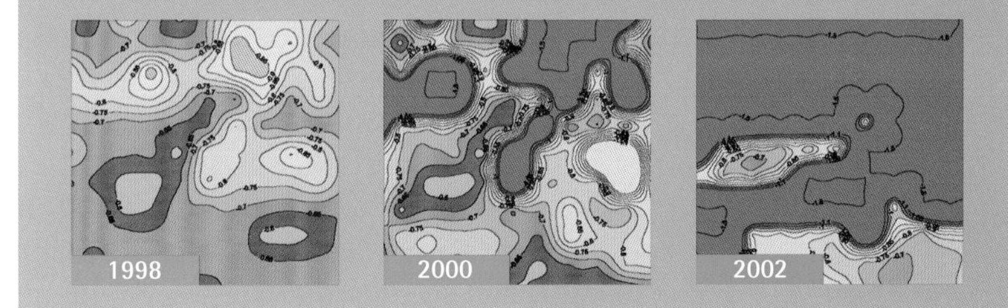

Seasonal Switch from Carbon Sink to Carbon Source

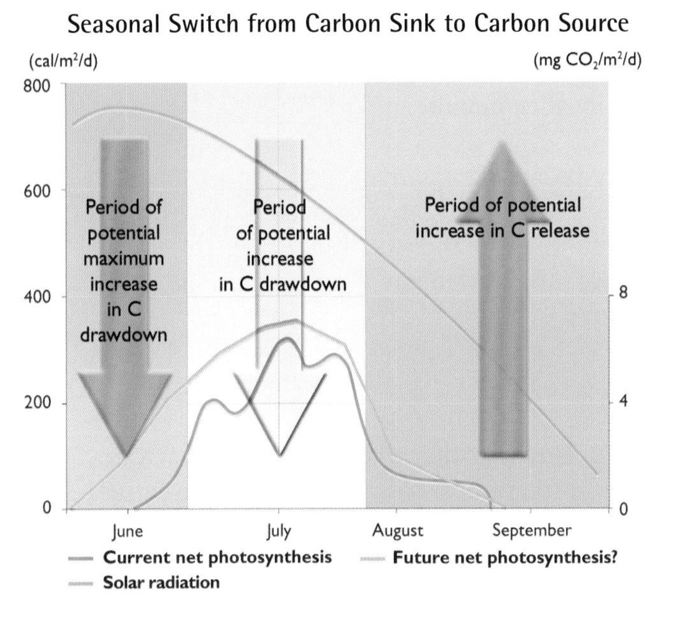

Changing Landscape Dynamics with Warming

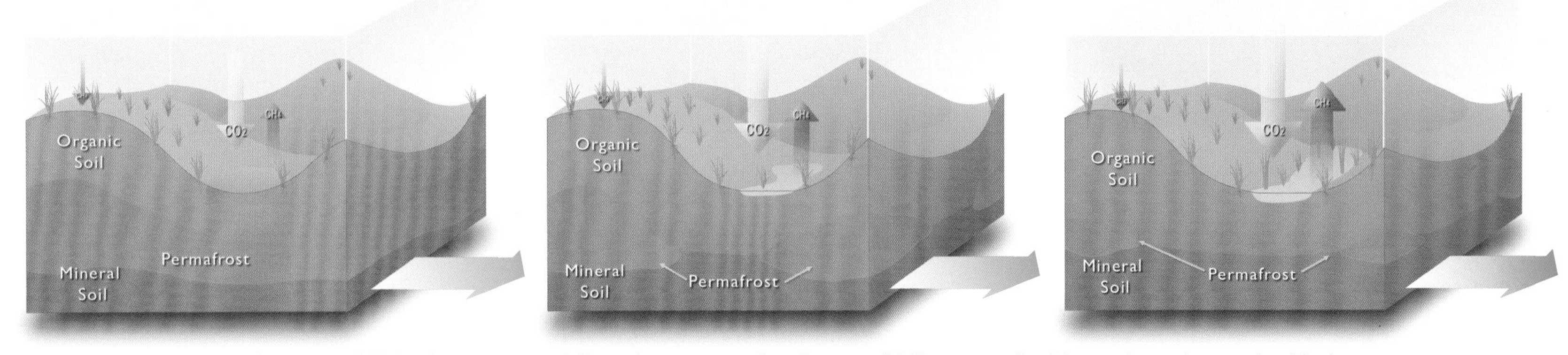

In northern Norway, Sweden, and Finland, many areas of discontinuous permafrost have small hills or mounds with wet depressions, each with characteristic vegetation (left). As climate warms, permafrost thaws and the wet areas increase in extent. The more productive vegetation captures more carbon dioxide but the greater extent of wet areas leads to greater methane emissions (middle) (this is already being observed). Eventually (right), the permafrost thaws completely, and the balance between methane emissions and carbon dioxide drawdown depends on subsequent drainage and precipitation.

3 Arctic vegetation zones are very likely to shift, causing wide-ranging impacts.

Northern Forests

Huge areas of central and eastern Siberia and northwestern North America represent the most extensive remaining areas of natural forest on the planet. Three of the four nations with the largest areas of forest in the world are arctic nations: Russia, Canada, and the United States. Forest and woodland areas in arctic nations represent about 31% of the world's forest (all types) and the boreal forest by itself covers about 17% of the earth's land surface area. As climate continues to warm, the boreal (northern) forests are projected to shift into the arctic region as the forested area expands northward.

The boreal forest collects, modifies, and distributes much of the freshwater that enters the arctic basin, and climate change will modify many of these important functions.

The boreal forests are extremely important globally for their economic and environmental values. Extensive areas of the boreal forests of Finland, Sweden, and parts of Canada are intensively managed for timber production and contribute 10-30% of the export earnings of these nations. The boreal forest collects, modifies, and distributes much of the freshwater that enters the arctic basin, and climate change will modify many of these important functions. The forest is also the breeding zone for a huge influx of migratory forest birds and provides habitat for fur-bearing mammals including wolverine, wolf, and lynx, as well as larger mammals including moose and caribou, all of which significantly support northern local economies.

Many impacts of climate change are already becoming apparent in the boreal forest: reduced rates of tree growth in some species and at some sites; increased growth rates in others; larger and more extensive fires and insect outbreaks; and a range of effects due to thawing permafrost, including new wetland development and collapsing of the ground surface and the associated loss of trees.

Challenges for Tree Establishment

Recent studies in Siberia have established conclusively that trees were present across the entire Russian Arctic, all the way to the northernmost shore, during the warm period that occurred about 8000-9000 years ago, a few thousand years after the end of the last ice age. Remains of frozen trees still in place on these lands provide clear evidence that a warmer arctic climate allowed trees to grow much further north than they are now. While this and other evidence suggests that vegetation zones are very likely to shift northward over the long term, such a process rarely develops in a straightforward manner. Various factors, including disturbances such as fires and floods, can either speed up or block tree establishment for some period of time. In addition, human activities create pressures that can prevent tree establishment in new areas. For example, in some parts of Russia, treeline is actually regressing southward due to the effects of industrial pollution.

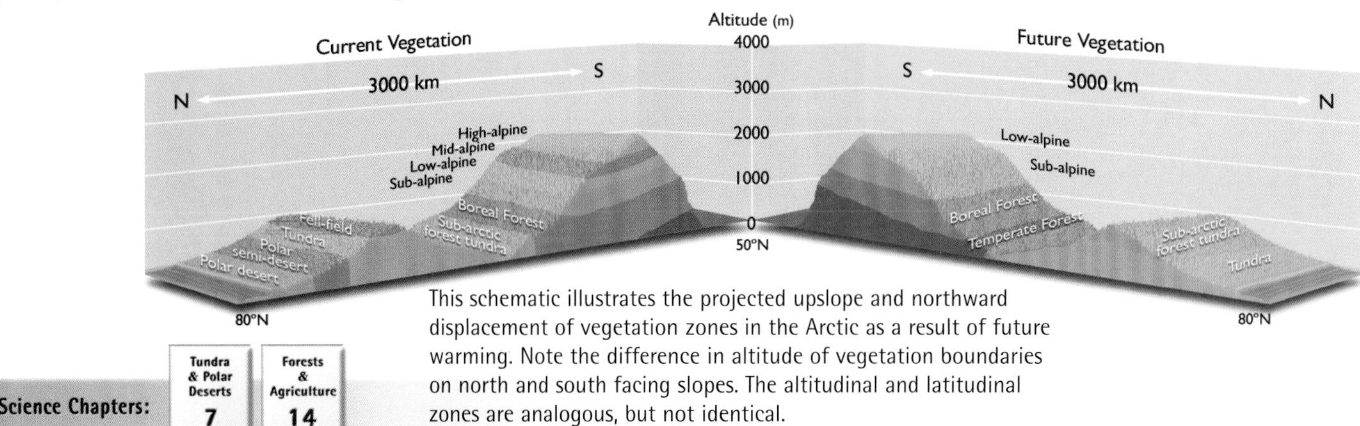

This schematic illustrates the projected upslope and northward displacement of vegetation zones in the Arctic as a result of future warming. Note the difference in altitude of vegetation boundaries on north and south facing slopes. The altitudinal and latitudinal zones are analogous, but not identical.

Science Chapters: Tundra & Polar Deserts 7 | Forests & Agriculture 14

Though forests are generally expected to move into tundra areas, some environments that now support trees will no longer be able to, primarily due to drying. New areas are likely to have a climate suitable for growing trees, but that does not guarantee that trees will, in fact, grow there, as there are a variety of challenges for trees moving into new areas. First, there is likely to be a time lag because some conditions needed for new tree growth, such as suitable soil, may not be present and will take time to develop. In addition, the dry tundra mat is not a hospitable surface for seeds to germinate in and plants to become established on. Some types of disturbance, such as flooding in river flood plains, are likely to facilitate tree establishment in some areas. On the other hand, in western Siberia, for example, it is possible that waterlogging of trees and a subsequent southward retreat of the treeline will accompany a wetter climate.

Disturbances such as fires and floods can either speed up or block tree establishment; human activities can also create pressures that block tree establishment in new areas.

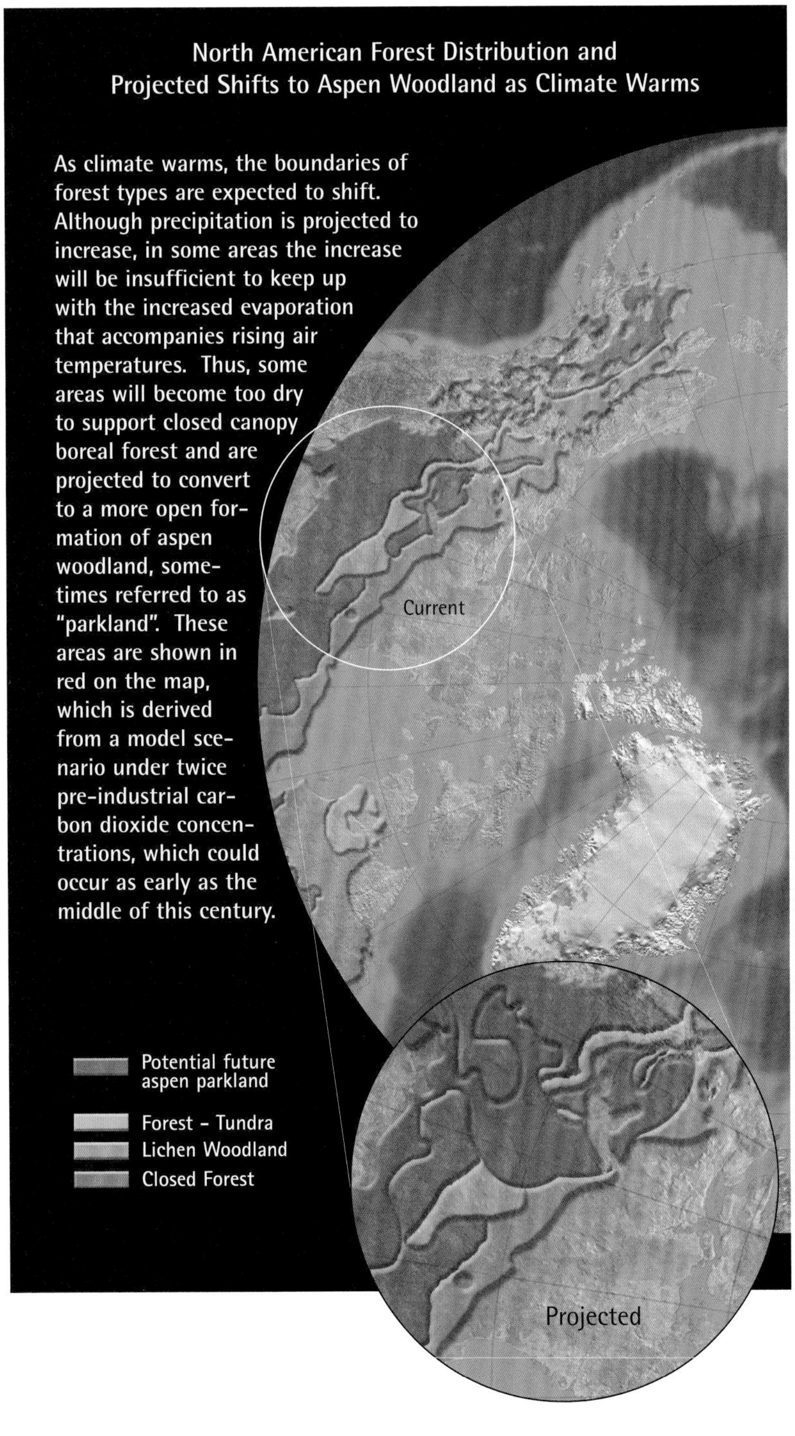

North American Forest Distribution and Projected Shifts to Aspen Woodland as Climate Warms

As climate warms, the boundaries of forest types are expected to shift. Although precipitation is projected to increase, in some areas the increase will be insufficient to keep up with the increased evaporation that accompanies rising air temperatures. Thus, some areas will become too dry to support closed canopy boreal forest and are projected to convert to a more open formation of aspen woodland, sometimes referred to as "parkland". These areas are shown in red on the map, which is derived from a model scenario under twice pre-industrial carbon dioxide concentrations, which could occur as early as the middle of this century.

Current

Projected

Potential future aspen parkland

Forest - Tundra

Lichen Woodland

Closed Forest

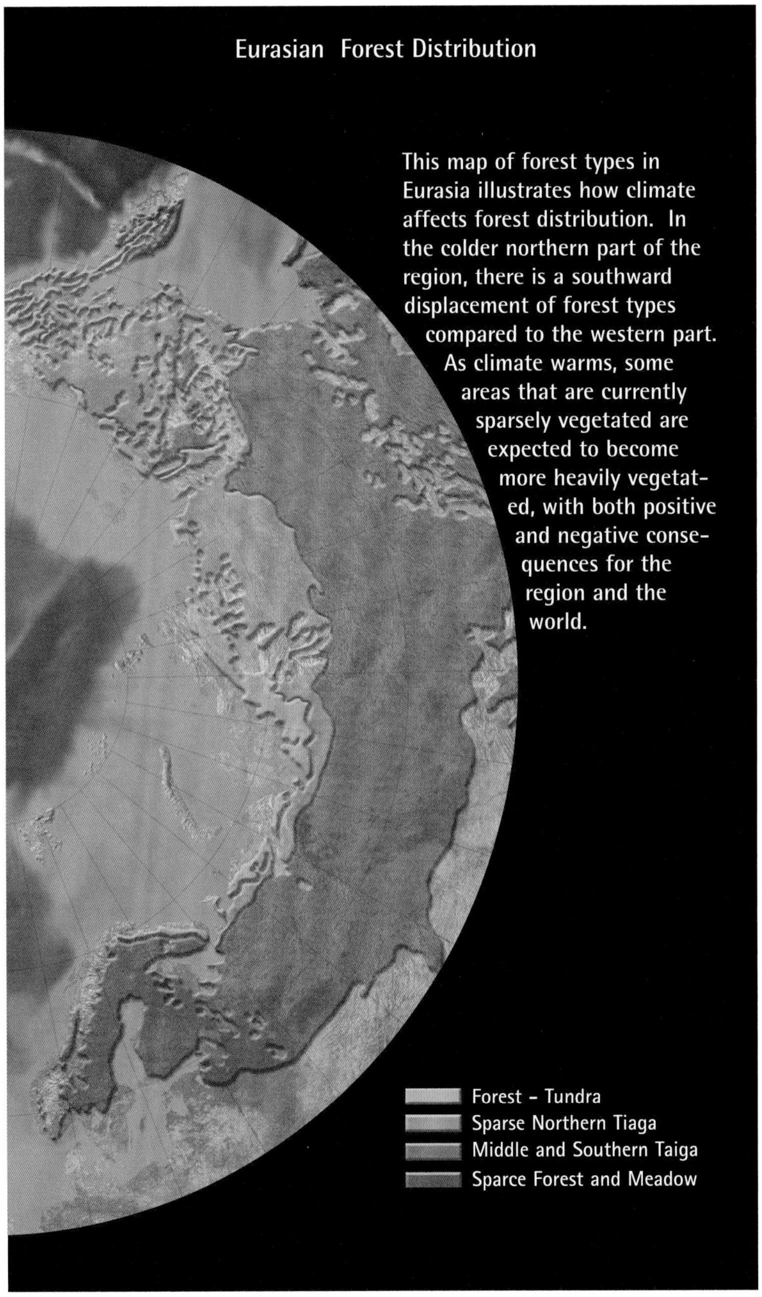

Eurasian Forest Distribution

This map of forest types in Eurasia illustrates how climate affects forest distribution. In the colder northern part of the region, there is a southward displacement of forest types compared to the western part. As climate warms, some areas that are currently sparsely vegetated are expected to become more heavily vegetated, with both positive and negative consequences for the region and the world.

Forest - Tundra

Sparse Northern Tiaga

Middle and Southern Taiga

Sparce Forest and Meadow

3 Arctic vegetation zones are very likely to shift, causing wide-ranging impacts.

Response of Siberian Forests to Climate Change

A forest study conducted in Siberia from the southern edge of the Central Asian steppe (grassland) to the treeline in the north reveals some of the ways that climate controls the growth of dominant tree species, in this case, the Scots pine and Siberian larch. In the southern part of this area, drought is the major factor limiting tree growth; cool wet growing seasons produce the most growth. Further north, in the southern and middle boreal forest, warmer midsummer weather reduces growth, while extension of the growing season earlier and later is associated with increased growth. In the northern boreal forest and northern treeline, warm midsummer weather is the principal factor that increases tree growth.

Sometimes, ecosystems display little change until they are confronted with environmental changes that exceed critical thresholds to which they are sensitive.

Climate change could produce two different kinds of response in the boreal forest. If the limiting environmental factors remain similar to those in the recent past, simple linear change would occur in which a forest type is replaced by its neighbor from the south. However, sometimes ecosystems display little change until they are confronted with environmental changes that exceed critical thresholds to which they are sensitive. In this case, climate change produces new kinds of ecosystems that are not present in the current landscape. Reconstructions of ancient ecosystems during past periods of climate change such as the last ice age, display such non-linear patterns of change. Potential non-linear changes might include a retreat of treeline southward in some areas. In other areas, forest tree species might not survive or might become so sparse that the tundra would directly border grassland or savanna rather than the boreal forest as it does today.

Siberian Larch and Warm Season Temperature

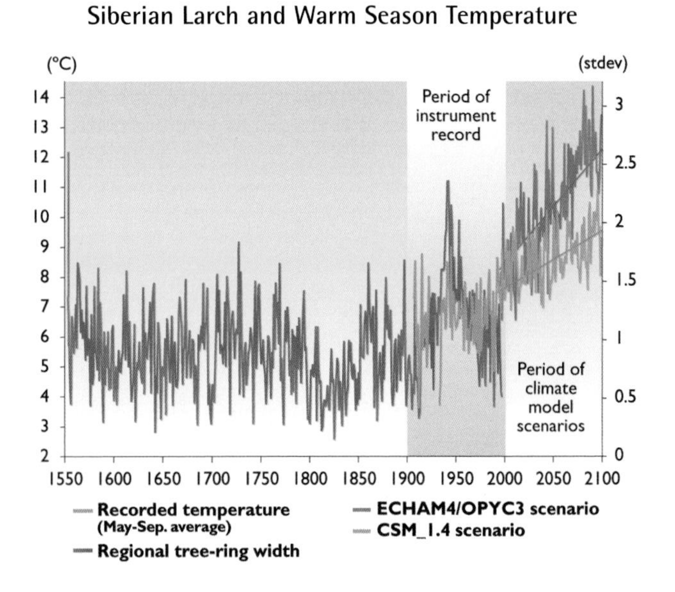

The graph shows the historical relationship between growth of Siberian larch and warm season temperature and two future warming scenarios in Russia's Taymir Peninsula. These trees respond positively to temperature increases. The warmer of the two scenarios above (ECHAM/OPYC3) would roughly double the growth rate and make this marginal site a productive forest. (The "site" is actually an average of four climate stations on the Taymir Peninsula.) The CSM_1.4 scenario would eliminate periods during which growth is severely limited by temperature.

White Spruce Response to Warming

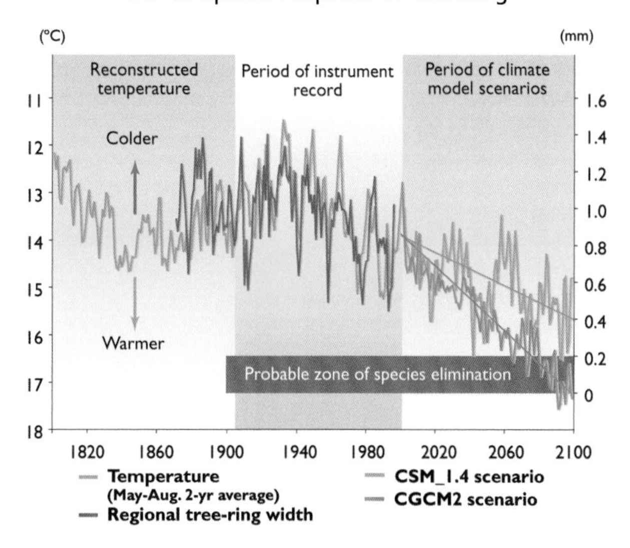

The graph shows the historic and projected relationship between white spruce growth and summer temperature in central Alaska. A critical temperature threshold was crossed in 1950, after which the growth began to fall. The projection of the Canadian climate model (CGCM2) suggests that this species is likely to be eliminated in the region by the latter part of this century.

Temperature Threshold for White Spruce

White spruce is the most widespread boreal conifer (evergreen, cone-bearing) and the most valuable timber species in the North American boreal forest. It also makes up most of the slow-growing forest found near the tundra margin. In dry central Alaska and western Canada, high summer temperatures decrease the growth of white spruce due to drought. By contrast, in moist coastal and lower mountain regions, white spruce growth increases with high summer temperatures. A study at treeline in the Alaska and Brookes Ranges examined 1500 white spruce trees in both dry and moist areas and found that 42% of the trees grew less under higher summer temperatures (negative responders) while 38% grew more (positive responders).

Most significantly, this study found a specific temperature threshold above which the growth of the negative responders decreased sharply. When July temperature at a nearby station rose above 16˚C, the growth of the negative responders decreased in direct proportion to the warming. Before 1950, few Julys were warmer than the threshold, so the negative response was weak. But since 1950, there have been many warm Julys, so the negative response has been very strong. Extending the observed pattern into the future, a 4˚C rise in July temperature would result in no growth, causing the elimination of these trees at treeline. (As for the positive responders, at most treeline sites in the Arctic, their positive response to warming has weakened in the late 20th century, although in Alaska their growth has increased with warming.)

The critical threshold of 16˚C is now frequently exceeded, strongly decreasing growth in this white spruce population. Extending the observed pattern into the future, a 4˚C rise in summer temperature would result in the elimination of these trees at treeline.

Specific temperature thresholds also trigger white spruce cone production, which is timed to release large numbers of seeds only when conditions are optimal for their establishment, generally following fires. Climate change has altered the timing of both fires and cone production so that these events are no longer as closely related, which could reduce the effectiveness of white spruce reproduction.

Black Spruce, Rising Temperatures, and Thawing Permafrost

Black spruce is the dominant tree in about 55% of the boreal forest in Alaska. It is a key species because its high absorption of the sun's energy increases warming and because of its role as a flammable species that carries fire across the landscape. While black spruce is likely to be an important part of the boreal forest that would expand into tundra as a result of climate change, the species faces a number of challenges to its survival where it is currently dominant. On drier permafrost-dominated sites in interior Alaska, the growth of black spruce decreases with increasing summer temperatures. At the upper ranges of projected warming for this century, it is not likely to survive on these sites due to drought conditions. On other sites in Alaska, black spruce is negatively affected by high early spring temperatures because photosynthesis (and thus the requirement for water) begins while the ground is still frozen, causing damage due to drying of its needles. Finally, even on those permafrost sites where warming has historically increased the growth of black spruce, the trees are at risk from collapsing of the ground surface due to thawing.

Black Spruce Response to Warming

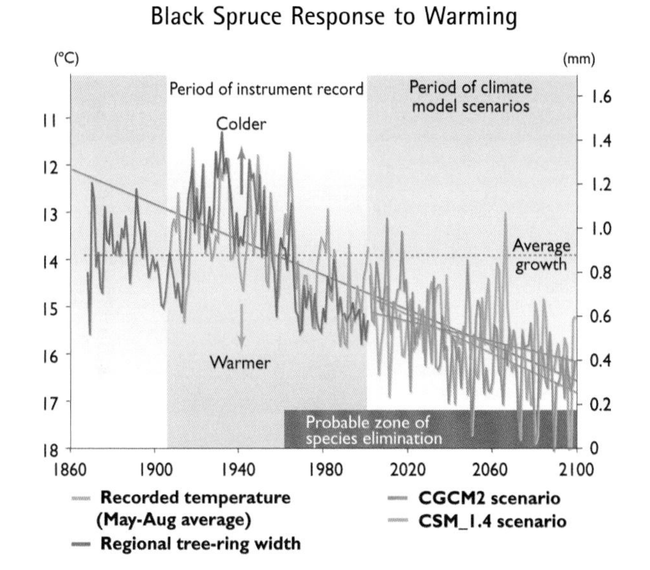

The graph shows the relationship of summer temperatures at Fairbanks, Alaska and relative growth of black spruce, historically and for two future warming scenarios. Average summer temperature is an excellent predictor of black spruce growth, with warm years resulting in strongly reduced growth. By 2100, temperatures projected by both scenarios would not allow the species to survive.

3 Arctic vegetation zones are very likely to shift, causing wide-ranging impacts.

Spruce Bark Beetle

Insect Outbreaks

Increased forest disturbances due to insect outbreaks are almost certain to result from climate warming. Increasing problems with spruce bark beetles and spruce budworms in the North American Arctic provides two important examples. Large areas of forest disturbance create new opportunities for invasive species from warmer climates and/or non-native species to become established.

Spruce Bark Beetle

The relationship of the spruce bark beetle to climate involves three factors, including two direct controls on insect populations and an indirect control on tree resistance. First, two successive cold winters depress the survival rate of the bark beetle to a level low enough that there is little outbreak potential the following summer. However, winters have been abnormally warm for decades in the North American Arctic, so the conditions for this control have not been met for some time. Second, the bark beetle normally requires two years to complete its life cycle, but in abnormally warm summers, it can complete its life cycle in one year, dramatically increasing the population and the resulting damage. This has occurred recently in Alaska and Canada.

Spruce Beetle Infestations in the Yukon, 1994-2002

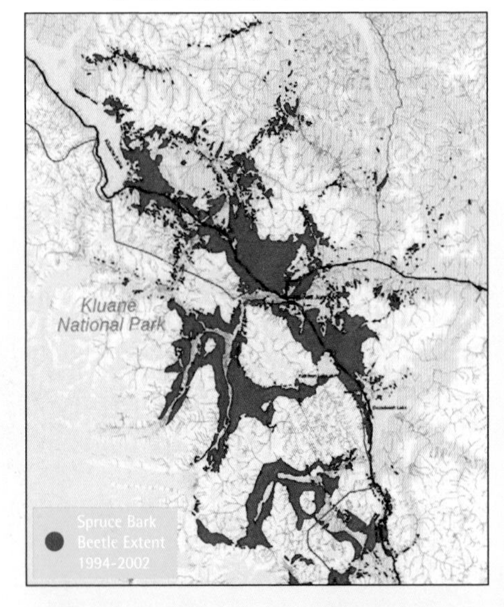

The spruce bark beetle has killed trees on about 300 000 hectares in the Alsek River corridor in Kluane National Park and in the Shakwak Valley north of Haines Junction since an outbreak was first identified in 1994. This is the largest and most intense outbreak of spruce bark beetle ever to affect Canadian trees. It is also the most northerly outbreak ever in Canada. 2002 was particularly intense, as aerial surveys recorded a 300% increase in the extent of infested areas as well as an increase in the severity of the attack.

Spruce Bark Beetle Outbreaks
Southern Kenai Peninsula

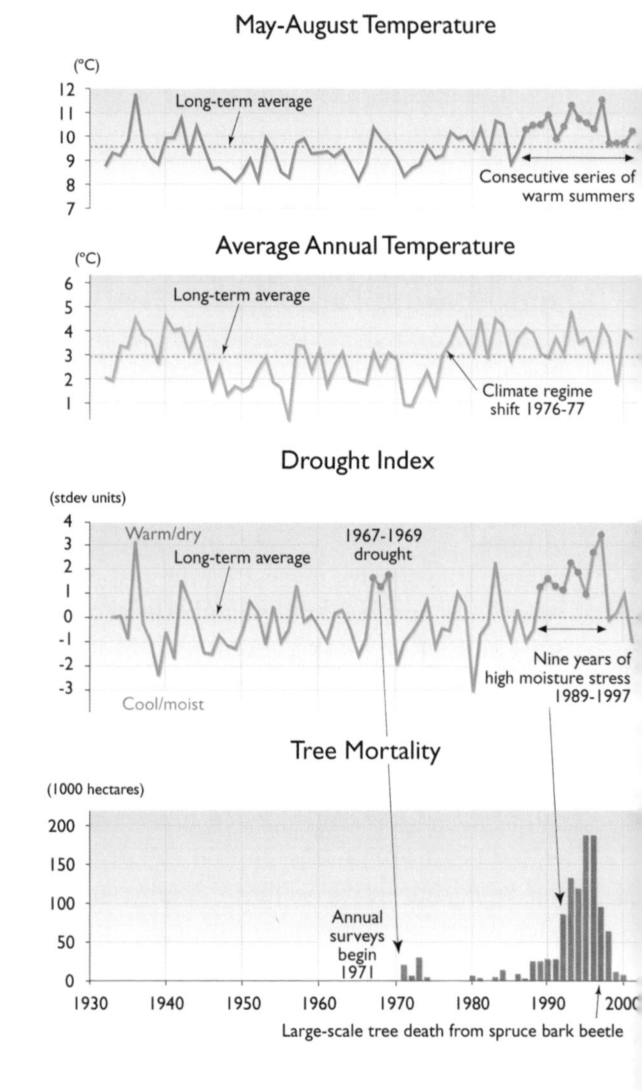

Large-scale tree death from spruce bark beetle

Spruce Beetle Activity Kenai Peninsula 1994-1999

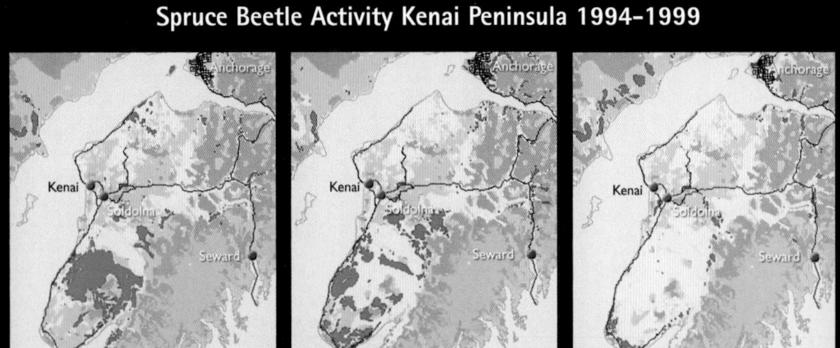

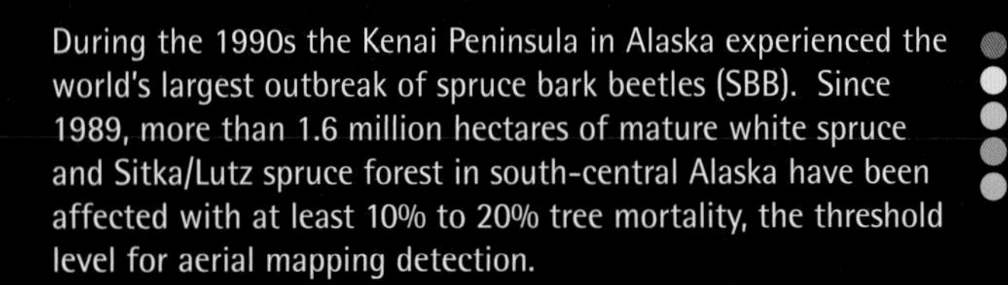

During the 1990s the Kenai Peninsula in Alaska experienced the world's largest outbreak of spruce bark beetles (SBB). Since 1989, more than 1.6 million hectares of mature white spruce and Sitka/Lutz spruce forest in south-central Alaska have been affected with at least 10% to 20% tree mortality, the threshold level for aerial mapping detection.

- Tree Mortality due to SBB
- Past Tree Mortality due to SBB
- Forest
- Non-Forest
- Glacier

Science Chapters: Past & Present Climate 2 | Forests & Agriculture 14

In addition, healthy spruce trees can successfully resist moderate numbers of beetle attacks by using their pitch, under high pressure, to push back against the female beetles trying to bore into the tree to lay eggs; the beetles are generally unable to overcome the flow of pitch. However, host trees under stress due to heat and drought have reduced growth reserves, leading to reduced amounts and lower pressure of pitch, and so a reduced ability to resist beetle attacks. When entire populations of trees are stressed by regional climatic events, such as has occurred recently in Alaska and parts of Canada, spruce bark beetle success is greatly increased and large-scale tree damage and loss occurs.

Spruce Budworm

Spruce Budworm

Weather is a critical factor in determining spruce budworm distribution. Sudden upsurges in budworm numbers generally follow drought and the visible effects of these outbreaks begin after hot, dry summers. Drought stresses the trees, reducing their resistance, and elevated summer temperatures increase budworm reproduction. For example, female budworms lay 50% more eggs at 25˚C than at 15˚C. Also, higher temperatures and drought can shift the timing of budworm reproduction such that their natural predators are no longer effective in limiting budworm numbers. Conversely, cold weather can stop a budworm outbreak. Budworms starve if a late spring frost kills the new shoot growth of the trees on which the larvae feed.

The entire range of white spruce forests in North America is considered vulnerable to outbreaks of spruce budworm under projected climate change.

Thus it is to be expected that climate warming would result in the northward movement of the spruce budworm and this has already occurred. Before 1990, spruce budworm had not appeared able to reproduce in the boreal forest of central Alaska. Then, in 1990, after a series of warm summers, a sudden and major upsurge in spruce budworm numbers occurred and visible damage to the forest canopy spread over several tens of thousands of hectares of white spruce forest. Populations of spruce budworm have since persisted in this area near the Arctic Circle. The entire range of white spruce forests in North America is considered vulnerable to outbreaks of spruce budworm under projected climate change. In the Northwest Territories of Canada, for example, the northern limit of current spruce budworm outbreaks is approximately 400 kilometers south of the northern limit of its host, the white spruce. Therefore, there is potential for a northward expansion of spruce budworm to take over this remaining 400 kilometer-wide band of currently unaffected white spruce forest.

Spruce Budworm infestation in Canada.

3 Arctic vegetation zones are very likely to shift, causing wide-ranging impacts.

Forest Fires

Fire is another major disturbance factor in the boreal forest and it exerts pervasive ecological effects. The area burned in western North America has doubled over the past thirty years, and it is forecast to increase by as much as 80% over the next 100 years under projected climate warming. Models of forest fire in parts of Siberia suggest that a summer temperature jump from 9.8°C to 15.3°C would double the number of years in which there are severe fires, increase the area of forest burned annually by nearly 150%, and decrease average wood stock by 10%.

Fires in Eurasian Forests

The area of boreal forest burned annually in Russia averaged four million hectares over the last three decades, and more than doubled in the 1990s. As climate continues to warm, the forest fire season will begin earlier and last longer. Projected changes in climate would greatly increase the area subjected to the types of weather that cause extreme fire danger. Under those conditions, fire spread in boreal forests of Eurasia would greatly increase if ignitions occurred. Fires are also expected to be more frequent and of higher intensity in all ecosystems, including bogged forests and peat bogs that that contain vast amounts of carbon-based forest materials such as moss, wood, and leaves. It is projected that about one billion tonnes of this organic matter will burn annually, increasing carbon emissions to the atmosphere. However, some scenarios suggest that it is possible that fire would become more frequent in some regions and less frequent in others, greatly affecting individual areas but amounting to only a small overall change in the total amount of fire.

About four million hectares of boreal forest in Russia burned annually over the past three decades, and the amount more than doubled in the 1990s.

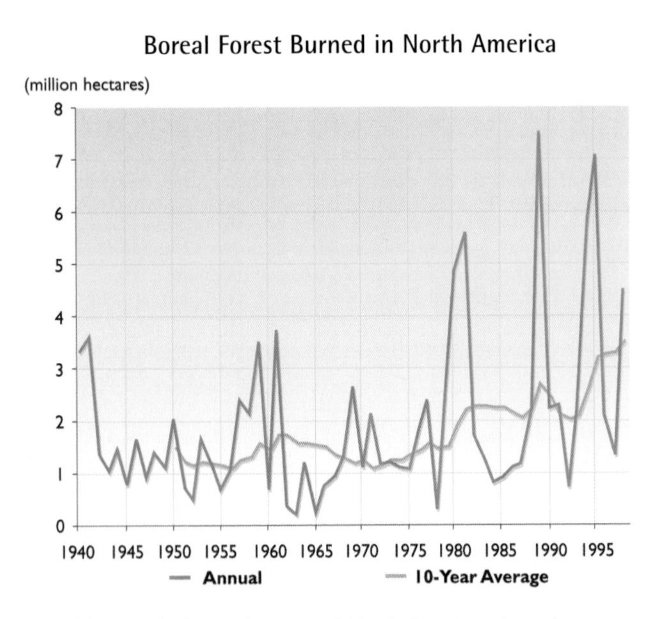

The graph shows the area of North American boreal forest that burned each year, in millions of hectares. The average area burned has more than doubled since 1970, coinciding with climatic warming in the region.

Projected Change in Growing Season Length by 2070-2090
Minimum Temperature greater than 0°C

(days) +70 +60 +50 +40 +30 +20 +10 0 -10 -20 -30 -40 -50 -60 -70

The colors indicate the change in the number of days in the growing season from the present to 2070-2090 under the Hadley 3 climate scenario. An average of three climate model's results suggests about a 20-30 day increase in the growing season for areas north of 60° latitude. The growing season is defined as the number of consecutive days in which the minimum temperature is above 0°C.

Agricultural Opportunities are Likely to Expand

Arctic agriculture is a relatively small enterprise in global terms, though some nations, such as Iceland, produce more than enough meat and dairy products to sustain their populations. Agriculture in the north consists mostly of cool-season forage crops; cool-season vegetables; small grains; raising cattle, sheep, goats, pigs, and poultry; and herding reindeer. While agriculture is limited by climate in the Arctic, especially in the cooler parts, it is also limited by the lack of infrastructure, small population base, remoteness from markets, and land ownership issues. Climatic limitations include short growing seasons (not enough time for crops to mature or to produce high yields of harvestable crops), lack of heat energy (days not warm enough during the growing season), long cold winters that can limit survival of many perennial crops, and moisture stress in some areas.

Climate warming is projected to advance the potential for commercial crop production northward throughout this century.

Climate change is projected to advance the potential for commercial crop production northward throughout this century, with some crops now suitable only for the warmer parts of the boreal region becoming suitable as far north as the Arctic Circle. Average annual yield potential is likely to increase as the climate becomes suitable for higher yielding varieties and the probability of low temperatures limiting growth is reduced. However, in warmer areas, increased warmth during the growing season may cause slight decreases in yields since higher temperatures speed development, reducing time to accumulate dry matter. Longer and warmer growing seasons are expected to increase the potential number of harvests and hence seasonal yields for perennial forage crops.

Uncertainty about winter conditions make forecasts about survival potential for crops difficult. Warmer winters could actually decrease survival of some perennial crops if winter thaws followed by cold weather become more frequent. This would be especially true in areas with little snowfall. However, longer growing seasons, especially in autumn, should result in a northward extension of conditions suitable for producing crops such as alfalfa and barley.

Water deficits are likely to increase over the next century in most of the boreal region because increases in warm-season precipitation are unlikely to keep up with increased evaporation due to higher temperatures. Unless irrigation is used, water stress is likely to negatively affect crop yields and water limitation is likely to become more important than temperature limitations for many crops in much of the region. Areas that are unlikely to experience water deficits include parts of eastern Canada, western Scandinavia, Iceland, and the Faroe Islands, which all experience fairly maritime climates.

Insects, diseases, and weeds are likely to increase throughout the Arctic with climate warming, although these problems are unlikely to offset potential yield increases or potential for new crops in most cases. However, severe outbreaks could indeed have that effect. For example, studies suggest that climate warming in Finland will increase the incidence of potato blight to the point that it will significantly decrease potato yield in that country.

Lack of infrastructure, small population (limited local markets), and long distance to large markets are likely to continue to be major factors limiting agricultural development in most of the Arctic during this century.

4 Animal species' diversity, ranges, and distrubution will change.

The Arctic is home to animal species that are admired around the world for their strength, beauty, and ability to survive in the harsh northern environment. Animals including caribou/reindeer, polar bears, and many species of fish and seals are also an essential part of the economy, diet, and culture for arctic peoples. Climate change will impact arctic species in ways that will affect conservation efforts as well as those who harvest wildlife resources on land and sea.

In the Marine Environment

More than half of the Arctic region is comprised of ocean. Many arctic life forms rely on productivity from the sea, which is highly climate-dependent. Climate variations have profound influences on marine animals. For example, the climate-related collapse of capelin in the Barents Sea in 1987 had a devastating effect on seabirds that breed in the area. And years with little or no ice in the Gulf of St. Lawrence in Canada (1967, 1981, 2000, 2001, 2002) resulted in years with virtually no surviving seal pups, when in other years, these numbered in the hundreds of thousands.

Polar bears are unlikely to survive as a species if there is an almost complete loss of summer sea-ice cover.

Polar Bears

Polar bears are dependent on sea ice, where they hunt ice-living seals and use ice corridors to move from one area to another. Pregnant females build their winter dens in areas with thick snow cover on land or on sea ice. When the females emerge from their dens with their cubs in spring, the mothers have not eaten for five to seven months. Their seal hunting success, which depends upon good spring ice conditions, is essential for the family's survival. Changes in ice extent and stability are thus of critical importance, and observed and projected declines in sea ice are very likely to have devastating consequences for the polar bear.

The earliest impacts of warming would be expected to occur at the southern limits of the bears' distribution, such as James and Hudson Bays in Canada, and such impacts have already been documented in recent years. The condition of adult polar bears has declined during the last two decades in the Hudson Bay area, as have the number of live births and the proportion of first-year cubs in the population. Polar bears in that region suffered 15% declines in both average weight and number of cubs born between 1981 and 1998. Later formation of sea ice in autumn and earlier break-up in spring means a longer period of annual fasting for female polar bears, and their reproductive success is tightly linked to their fat stores. Females in poor condition have smaller litters and smaller cubs that are less likely to survive. Climate change is also likely to increase bear deaths directly. For example, increased frequency and intensity of spring rains is already causing some dens to collapse, resulting in the death of females and cubs. Earlier spring break-up of ice could separate traditional den sites from spring feeding areas, and young cubs cannot swim long distances from dens to feeding areas.

Polar bears are unlikely to survive as a species if there is an almost complete loss of summer sea-ice cover, which is projected to occur before the end of this century by some climate models. The only foreseeable option that polar bears would have is to adapt to a land-based summer lifestyle, but competition, risk of hybridization with brown and grizzly bears, and increased human interactions would then present additional threats to their survival as a species. The loss of polar bears is likely to have significant and rapid consequences for the ecosystems that they currently occupy.

Ice-dependent Seals

Ice-dependent seals, including the ringed seal, ribbon seal, and bearded seal, are particularly vulnerable to the observed and projected reductions in arctic sea ice because they give birth to and nurse their pups on the ice and use it as a resting platform. They also forage near the ice edge and under the ice. Ringed seals are likely to be the most highly affected species of seal because all aspects of their lives are tied to sea ice. They require sufficient snow cover to construct lairs and the sea ice must be stable enough in the spring to successfully rear young. Earlier ice break-up could result in premature separation of mothers and pups, leading to higher death rates among newborns.

Adapting to life on land in the absence of summer sea ice seems highly unlikely for the ringed seal as they rarely, if ever, come onto land. Hauling themselves out on land to rest would be a dramatic change to the species' behavior. Giving birth to their pups on land would expose newborns to a much higher risk of being killed by predators. Other ice-dependent seals that are likely to suffer as sea ice declines include the spotted seal, which breeds exclusively at the ice edge in the Bering Sea in spring, and the harp seal, which lives associated with sea ice all year. Unlike these ice-associated seal species, harbour seals and grey seals are more temperate species with sufficiently broad niches that they are likely to expand their ranges in an Arctic that has less ice coverage.

Seabirds

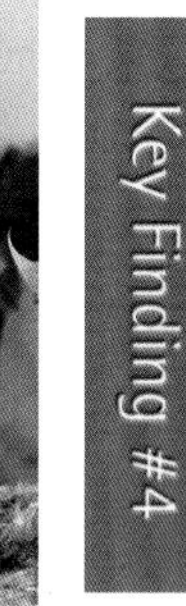

Some seabirds such as ivory gulls and little auks are very likely to be negatively impacted by the decline of sea ice and subsequent changes to the communities in which they live. The ivory gull is intimately associated with sea ice for most of its life, nesting and breeding on rocky cliffs that offer protection from predators, and flying to the nearby sea ice to fish through cracks in the ice and scavenge on top of the ice. As the sea ice edge retreats further and further from suitable coastal nesting sites, serious consequences are very likely to result. Major declines have already been observed in ivory gull populations, including an estimated 90% reduction in Canada over the past 20 years.

The Walrus and the Ice Edge

The ice edge is an extremely productive area and is very sensitive to climate change. The most productive areas are nearest the coasts, over the continental shelves. As sea ice retreats farther from the shorelines, the marine system will lose some of its most productive areas. For walrus in many areas, the ice edge provides the ideal location for resting and feeding because walrus are bottom feeders that eat clams and other shellfish on the continental shelves. As the ice edge retreats away from the shelves to deeper areas, there will be no clams nearby. Walrus also normally travel long distances on floating ice, allowing them to feed over a wide area.

4 Animal species' diversity, ranges, and distrubution will change.

Diver sampling sea-ice algae at Cape Evans. The underside of the sea ice is colored brown by the algae. Brine channels that form as the ice melts form pinnacles of ice hanging down into the water column and these become heavily colonized by ice algae.

Ice Algae and the Related Food Web

The vast reduction in multiyear ice in the Arctic Ocean is likely to be immensely disruptive to microscopic life forms associated with the ice, as they will lack a permanent habitat. Research in the Beaufort Sea suggests that ice algae at the base of the marine food web may have already been profoundly affected by warming over the last few decades. Results indicate that most of the larger marine algae under the ice at this site died out between the 1970s and the late 1990s, and were replaced by less-productive species of algae usually associated with freshwater. Researchers say that this is likely to be related to the fact that melting has formed a 30-meter thick layer of relatively fresh water below the remaining ice, one third deeper than it was 20 years before. Among the areas likely to be most severely affected by such changes will be the Bering Sea and Hudson Bay, in the lower Arctic, where sea ice is already disappearing earlier in spring and forming later in the autumn. As the Arctic continues to warm, sea ice will melt rapidly in the spring over continental shelf areas and withdraw toward the deep ocean of the central Arctic.

Research in the Beaufort Sea suggests that ice algae at the base of the marine food web may have already been profoundly affected by warming over the last few decades.

Additional Climate-Related Threats to Marine Species

Climate change poses risks to arctic marine mammals and some seabirds beyond the loss of habitat and forage bases. These include increased risk of disease due to a warmer climate, increased pollution impacts as rising precipitation brings more atmospheric and river-borne pollution northward, increased competition as temperate species expand their ranges northward, and impacts due to increased human traffic and development in previously inaccessible, ice-covered areas.

Arctic Marine Food Web

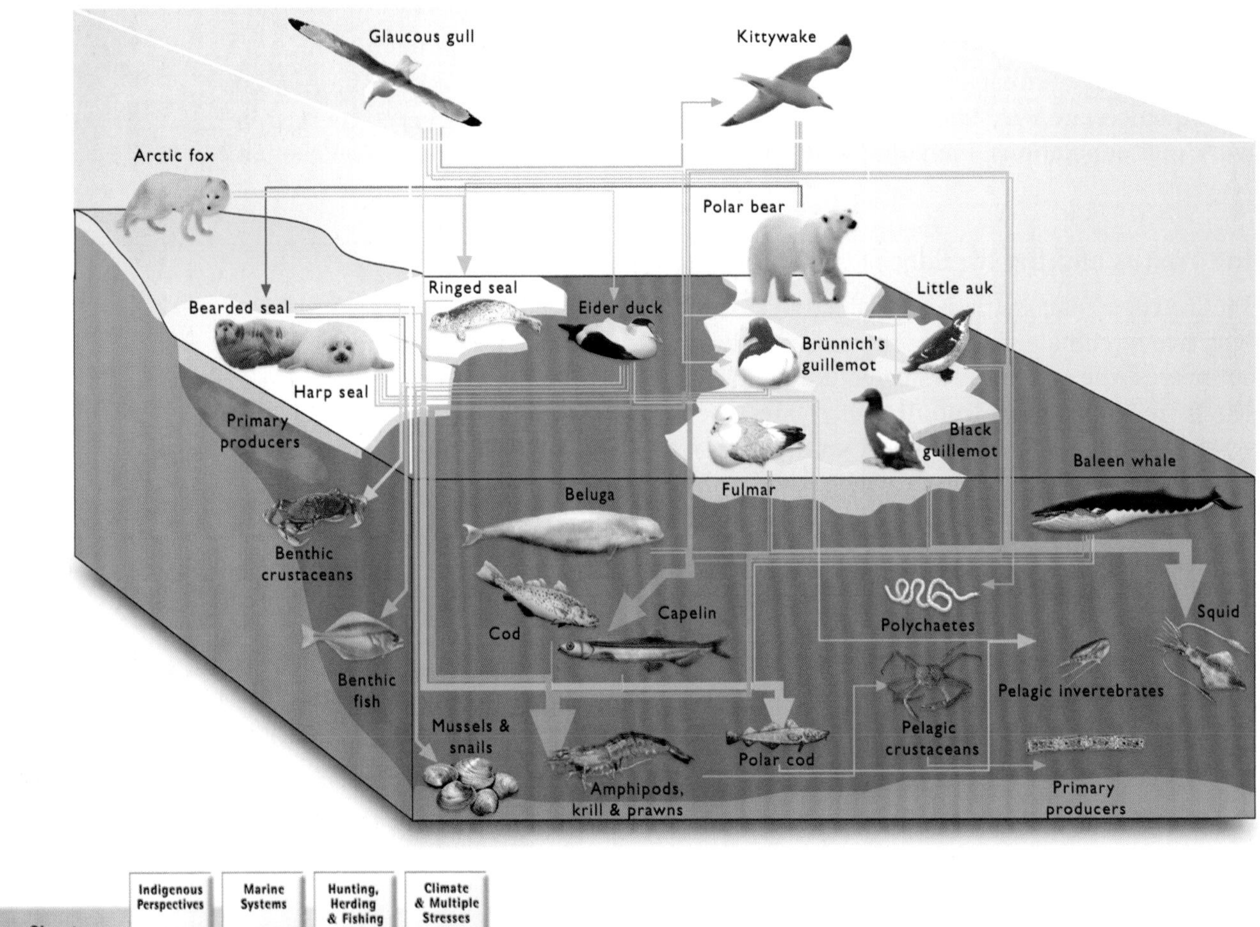

Science Chapters: Indigenous Perspectives 3 | Marine Systems 9 | Hunting, Herding & Fishing 12 | Climate & Multiple Stresses 17

Chemicals and Climate Interact to Impact Polar Bears

The increase in environmental stress on polar bears caused by climate change interacts with the stresses caused by chemical contaminants. Polar bears, at the top of the marine food chain, accumulate contaminants in their fat by eating ringed seals and other marine mammals who have absorbed the chemicals by eating contaminated species lower on the food chain. High levels of chlorinated compounds and heavy metals have been found in polar bears. In some cases, contaminants may be stored in fat, keeping the chemicals from affecting the bears' health when fat reserves are high. But during a poor feeding season, when the fat reserves must be used, the chemicals are released into the body. Polar bears in some areas of the Arctic have been observed to have less fat reserves in recent decades as sea ice breakup occurs progressively earlier, forcing them ashore where they are required to fast for increasingly longer periods.

Climate and Social Changes Interact to Impact Marine Hunters

Many arctic communities depend on hunting polar bear, walrus, seals, whales, seabirds, and other marine animals. Changes in the species' ranges and availability, and the decreased ability to travel safely in changing and unpredictable ice conditions are making people feel like strangers in their own land. Some societal changes have increased vulnerability to climate-induced changes. For example, over recent decades, many Inuit hunters have switched from dog sleds to snowmobiles, and while dogs could sense dangerous ice conditions, snowmobiles cannot. (On the other hand, snowmobiles allow people to hunt over larger areas and to transport bigger loads.) In addition, people are no longer nomadic, following animals' seasonal movements. Because people now live in permanent settlements, their ability to adapt to changing climatic conditions and/or animal availability by moving has been greatly reduced.

4 Animal species' diversity, ranges, and distrubution will change.

Marine Fisheries

Arctic marine fisheries provide an important food source globally, and a vital part of the economy of the region. Because they are largely controlled by factors such as local weather conditions, ecosystem dynamics, and management decisions, projecting the impacts of climate change on marine fish stocks is problematic. There is some chance that climate change will induce major ecosystem shifts in some areas that would result in radical changes in species composition with unknown consequences. Barring such shifts, moderate warming is likely to improve conditions for some important fish stocks such as cod and herring, as higher temperatures and reduced ice cover could possibly increase productivity of their prey and provide more extensive habitat.

Greenlandic Cod and Climate

A striking example of a positive climate-related impact involves West Greenland cod. Under the very cold conditions between around 1900 and 1920, there were few cod around Greenland. In 1922 and 1924, large numbers of cod were spawned in Icelandic waters and drifted from Iceland to East Greenland and then to West Greenland where they flourished, resulting in the start of a significant fishery beginning in the mid- to late 1920s. Large numbers of these cod returned to Iceland to spawn in the early 1930s and then remained there. However, many other cod stayed and spawned off West Greenland, giving rise to an independent, self-sustaining cod stock. During the warm period that spanned the middle of the 20th century, the Greenland cod stock grew very large, sustaining an annual average catch of about 315 000 tonnes between 1951 and 1970. The cold conditions that have prevailed since about 1965 appear to have rendered cod incapable of reproducing in Greenlandic waters. The only significant catches since then have been based on fish born in Icelandic waters in 1973 and 1983 that drifted to Greenland from Iceland.

While projected conditions are likely to benefit some species, such as cod, they are likely to negatively affect others, such as northern shrimp, necessitating adjustments in commercial fishery operations.

While projected conditions are likely to benefit some species, such as cod, they are likely to negatively affect others, such as northern shrimp, necessitating adjustments in commercial fishery operations. The area inhabited by some arctic species, including northern shrimp, will probably contract and the abundance of those species decrease. This would reduce the large catch (about 100 000 tonnes a year) of northern shrimp

Observed and Projected Harvests

Mixing Waters and Fish Drift Routes

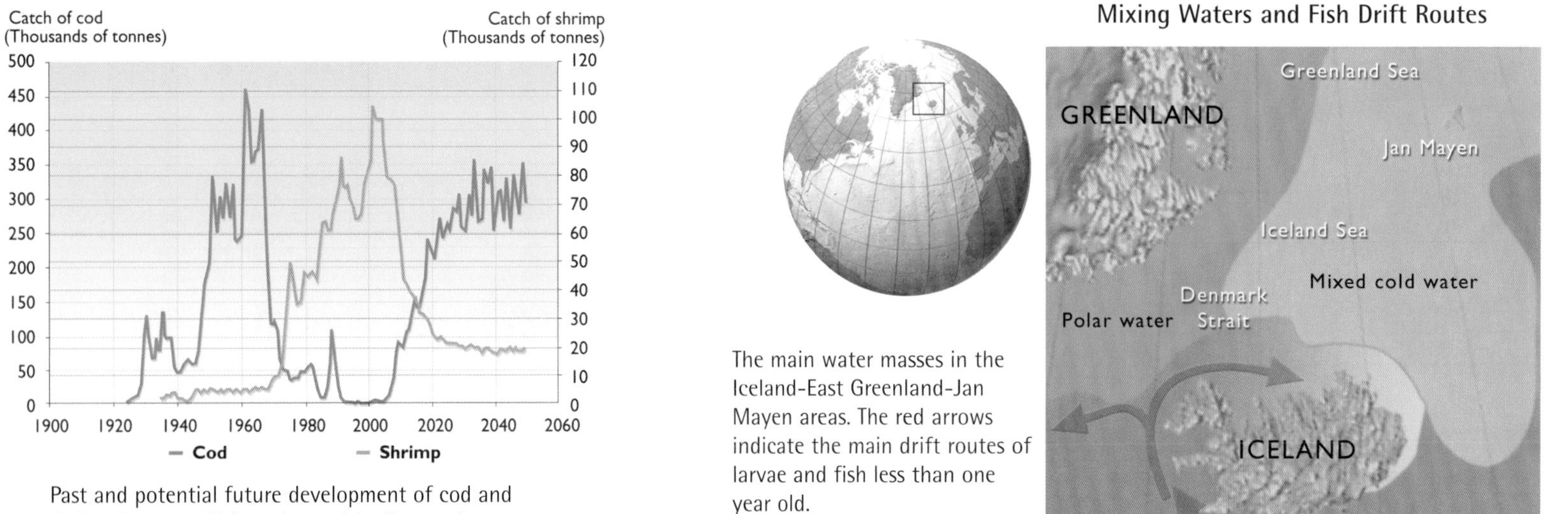

Past and potential future development of cod and shrimp harvest off Greenland with climate change.

The main water masses in the Iceland-East Greenland-Jan Mayen areas. The red arrows indicate the main drift routes of larvae and fish less than one year old.

currently taken from Greenlandic waters. Furthermore, northern shrimp are an important part of the diet of cod in the waters off Greenland. Thus, if the cod stock were to grow as it did in the last century, the decline in the northern shrimp population could negatively affect the diet and growth of the cod stock. Because the commercial value of a healthy cod stock would be much greater than the value of the shrimp catch, the shrimp fishery would have to be curtailed even further.

Climate, Overfishing, and Norwegian Herring

In the early 1950s, the stock of Norwegian spring-spawning herring was as large as 14 million tonnes, the world's largest herring stock, and was important to Norway, Iceland, Russia, and the Faroe Islands. At that time, these herring migrated west across the Norwegian Sea to feed in the zooplankton-rich waters north and east of Iceland as well as in the oceanic area between Iceland and the island of Jan Mayen (71°N, 8°W). In 1965, a sudden and severe cooling of these waters resulted in the decimation of the tiny crustacean (*Calanus finmarchicus*) that was by far the most important single food item in the diet of these herring. The herring's feeding areas were displaced to the east and northeast by several hundred nautical miles, thus placing the stock under severe environmental stress. In the 1960s, the stock was also subjected to severe overfishing and collapsed during the latter half of the decade. Although high fishing intensity of both adults and juveniles was the primary reason for the collapse, the climatic cooling probably contributed to the decline.

Herring Spawning Stock and Temperature

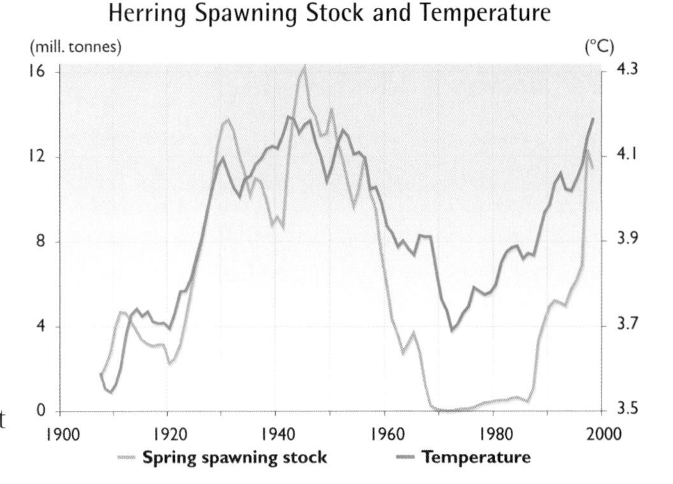

Norwegian spring spawning herring stocks increased greatly during the warming period of the 1920s-1930s and then declined rapidly beginning in the late 1950s. Overfishing was the primary cause of the collapse of the population, although climatic cooling was probably a contributing factor.

In the 1970s, the small numbers of herring that were left did not need to search far to feed and thus stayed close to the Norwegian shore. What was left of the fishery was strictly regulated, and fishing was prohibited for several years. These restrictions, coupled with favorable climatic conditions, contributed to the stock's increase to three to four million tonnes and limited fishing began again. In 1995, the stock reached five million tonnes and extended its feeding grounds and migratory range into international waters. The stock therefore became available for fishing outside Norway's jurisdiction, making the Norwegian management regime insufficient to protect the stock and threatening its continued recovery. In 1996, an agreement was reached between Norway, Russia, Iceland, the Faroe Islands, and the European Union to set quotas for allowable catches of Norwegian spring spawning herring. Such agreements will be crucial in the future as climate change alters fish stocks and their ranges.

Historic Changes in Migration Routes

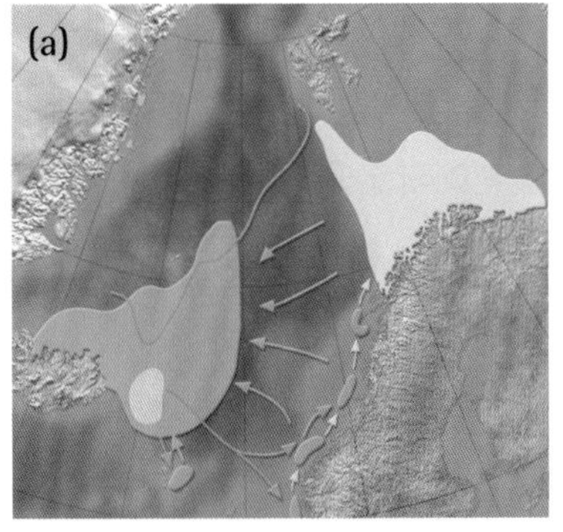

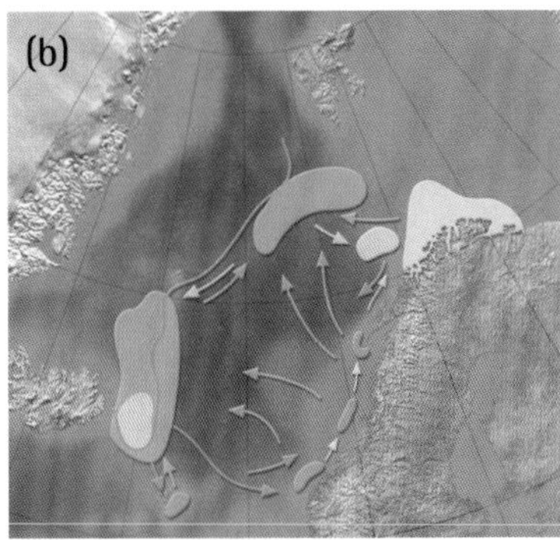

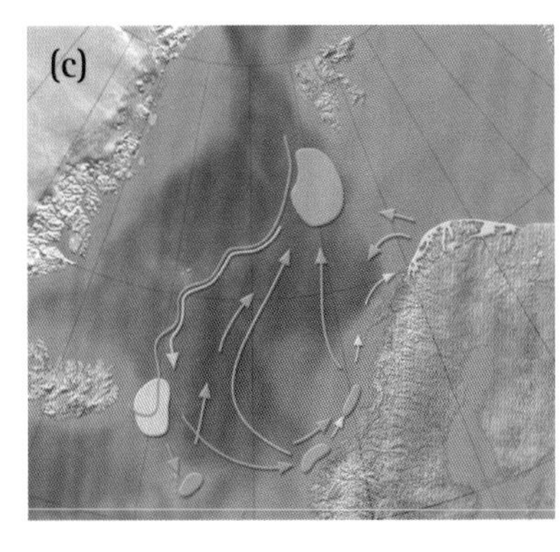

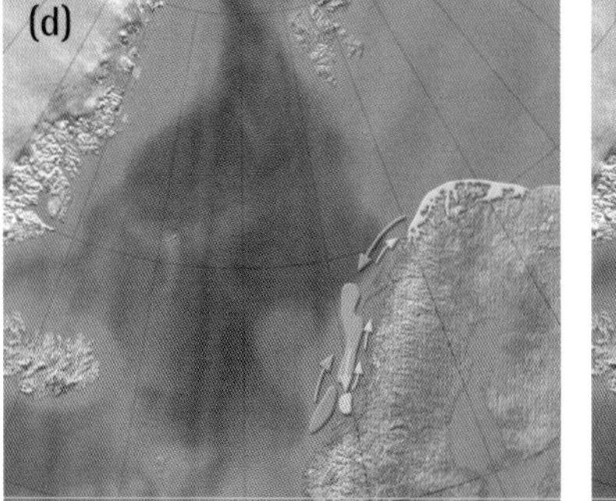

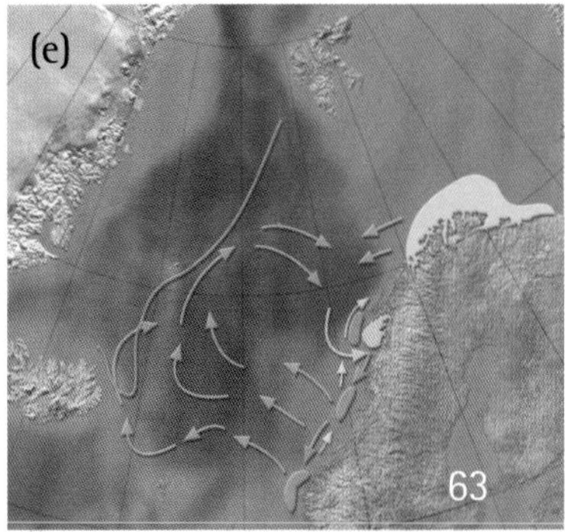

Changes of migration routes, and feeding and wintering areas of Norwegian spring spawning herring during the latter half of the 20th century. (a) Normal migration pattern during the warm period before 1965. (b-c) After a pulse of sea ice and freshwater from the Arctic sent cold, low-salinity water into the East Greenland and the East Icelandic currents, until the stock collapsed in 1968. (d) During years of low stock abundance (1972-1986). (e) The present day migration pattern.

- Spawning areas
- Juvenile areas
- Main feeding areas
- Spawning migrations
- Feeding migrations
- Spawning migrations

4 Animal species' diversity, ranges, and distrubution will change.

Climate Shifts and Fisheries Impacts

A climate shift occurred in the Bering Sea in 1977, abruptly changing from a cool to a warm period, perhaps a reflection of the Pacific Decadal Oscillation. The warming brought about ecosystem shifts that favored herring stocks and enhanced productivity for Pacific cod, skates, flatfish, and non-crustacean invertebrates. The species composition of living things on the ocean floor changed from being crab-dominated to a more diverse mix of starfish, sponges, and other life forms. Historically high commercial catches of Pacific salmon occurred. The Walleye pollock catch, which was at low levels in the 1960s and 1970s (two to six million tonnes), has increased to levels greater than ten million tonnes for most of the years since 1980.

The Bering Sea is experiencing a major warming in bottom water temperature that is forcing cold-water species of fish and mammals northward and/or into decline.

For most of the North Atlantic, the total effect of climate change on arctic and sub-arctic fish stocks is likely to be of lesser magnitude than the effects of fisheries management, at least for the next two to three decades. This is mainly due to the relatively small warming expected for the first part of the 21st century in this area. In the Bering Sea, however, rapid climate change is already apparent, and its impacts significant. The Bering Sea is experiencing a major warming in bottom water temperature that is forcing cold-water species of fish and mammals northward and/or into decline. The first concern of Bering Sea fisheries management is thus likely to be managing for the ecosystem reorganization that is and will continue to be taking place as a result of climate change.

While it seems unlikely that climate change effects on fisheries will have long-term arctic-wide social and economic impacts, certain areas that are heavily dependent on fisheries are likely to be affected. Very severe dislocations are possible and have occurred historically. For example, when the Labrador/Newfoundland cod fishery collapsed due to overfishing, shifts in oceanic conditions, and other factors in the early 1990s, many cod fishermen went out of business or switched to other species, and the value of the fish catch in the province declined sharply. The cod stock has still not recovered, although a decade has passed. The shrimp and crab fisheries that eventually replaced the cod fishery are much less labor intensive and employ far fewer people, although the total commercial value of the fishery is about twice the value of the cod fishery. So while the fishing industry can generally adapt at the national level, particular people and places can be strongly affected.

Eastern Bering Sea Catch, 1954-2000

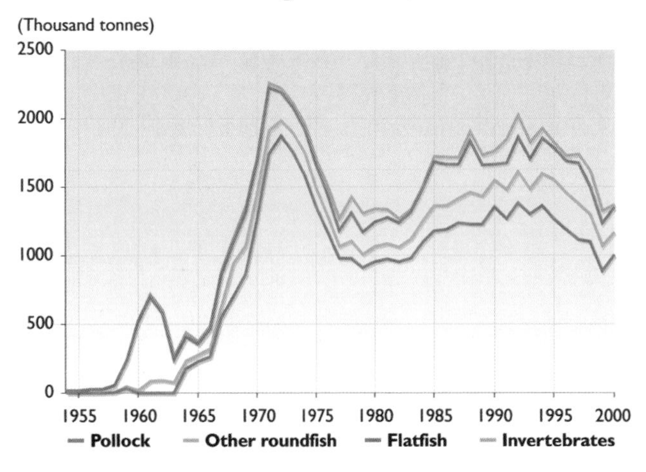

Western Bering Sea Catch, 1965-2001

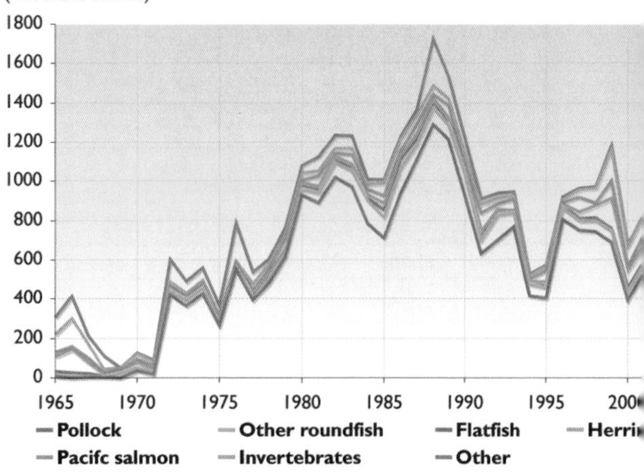

Seal Hunting, Fishing, and Climate Change in West Greenland: A Historical Perspective

Historical changes in West Greenland provide a good example of the relationship between climate change and associated social and economic changes. A climatic variation resulted in warming of the waters to the south and west of Greenland in the 1920s and 1930s, causing seal populations to shift northward, making seal hunting more difficult for the local Inuit. At the same time, cod (as well as halibut and shrimp) moved into the warmer waters, enabling the development of a cod fishery. Some local people, such as those in the west coast town of Sisimiut, were able to take advantage of the opportunities that arose due to social and technological factors. Sisimiut became an important fishing center with other new industries and a diverse economic base.

This stands in contrast to the development of the southwest Greenlandic town of Paamiut around the same time. Paamiut's development was based largely on plentiful resources of cod. With few other resources available in commercially viable quantities, there was little incentive to diversify the local economy. The concentration on a single resource made the town vulnerable to environmental change. When the cod population began to fall, due to a combination of climatic change and overfishing, the economy and population of Paamiut declined as a result. This points to the importance of recognizing in any adaptive strategy that local conditions (environmental, social, economic, technological, etc.) are important factors in determining the success of a region subject to change.

Possible Changes in Fish Distribution

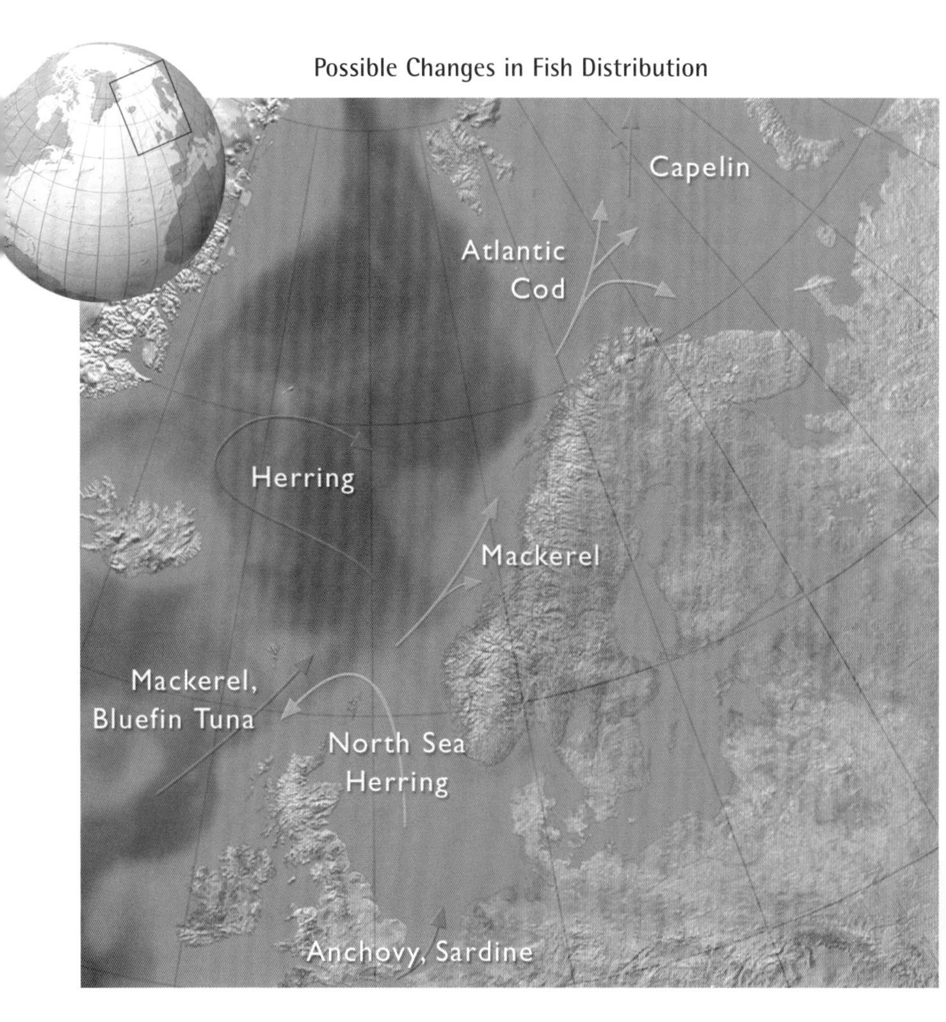

Possible changes in the distribution of selected fish species the Norwegian and Barents Seas resulting from an increase in ocean temperature of 1 to 2°C.

4 Animal species' diversity, ranges, and distrubution will change.

If the Atlantic waters that run north along the coast of Norway warm by a couple of degrees, aquaculture operations would probably have to shift northward, incurring significant costs.

Aquaculture

Salmon and trout are the two major aquaculture species in the Arctic, farmed by a high tech industry employing advanced equipment, in many ways more akin to pig or poultry farming than to fishing. Norway has developed a large industry over the past two decades and is now the world's largest farmed salmon producer. Total production in 2000 was valued at 1.6 billion US dollars, making salmon the single most important species in terms of economic value in the Norwegian fishing industry.

Slightly warmer water might be expected to increase fish growth rates but any more than slight warming could exceed the temperature tolerances of farmed species. In addition, warmer waters would have other negative effects, such as an increase in diseases and toxic algal blooms. If the Atlantic waters that run north along the coast of Norway warm by a couple of degrees, aquaculture operations would probably have to shift northward, incurring significant costs. Aquaculture in marine systems off Newfoundland and Labrador is problematic due to their latitudes. It is not uncommon for the temperature in the upper water layers to rise above the tolerance of many of the species currently cultivated.

The aquaculture industry depends on a huge supply of wild fish captured from the open ocean to provide the fishmeal and oils that are important dietary components of farmed fish such as salmon and trout. The quantity of supplies needed is so high that the industry is sensitive to rapid fluctuations in important wild fish stocks, and such fluctuations can be brought about by climatic factors. For example, El Niño events in the Pacific Ocean have already affected the industry through huge impacts on anchovy stocks. From 1997 to 1998, the global anchovy fishery was reduced by nearly eight million tonnes, mainly due to El Niño. Furthermore, many species now harvested elsewhere to provide fodder for farmed fish are also highly important in the diet of wild stocks that are of much greater commercial value but which are presently not abundant

due to overfishing. Should fisheries managers succeed in enlarging these wild stocks, large reductions might be needed in the fish farming operations that currently turn these important prey species into fishmeal and oil.

Aquaculture in the Faroe Islands

The ocean surrounding the Faroe Islands is part of the most important feeding grounds for wild stocks of northern European Atlantic salmon. The islands of this archipelago are loosely clustered, and have short fjords and inlets, creating a relatively open area with pronounced ocean currents that prevent water stagnation. This offers good conditions for farming Atlantic salmon and rainbow trout, by far the most prominent farmed species. In the 1980s, fish farming became an industry in the Faroe Islands, with annual production reaching about 8000 tonnes by 1988. In the early-to-mid 1990s, the industry collapsed because the numerous small fish farms could not survive the large drop in the market price of farmed salmon. Fish diseases also played a role in the collapse. Production increased again in the late 1990s, and in 2001, Faroese fish farming was consolidated into a few large companies that now operate at 23 sites. There is now a fish farm in nearly every suitable bay and fjord in the archipelago.

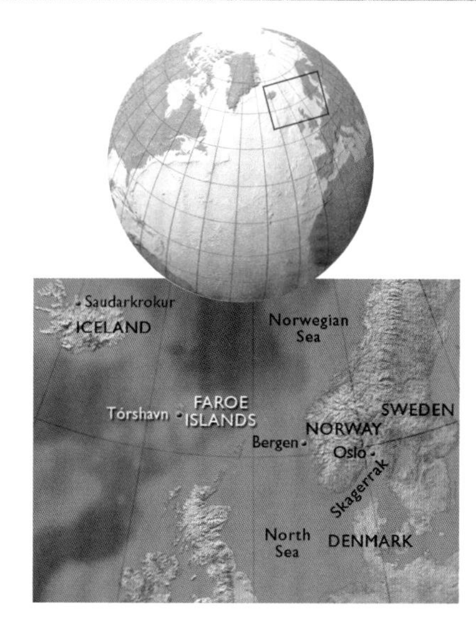

The Faroe Islands have become an important international player in salmon farming, harvesting a record 53 000 tonnes (gutted weight) of salmon and rainbow trout, valued at about 180 million US dollars in 2003. With a population of approximately 45 000 people, this corresponds to a production of almost 1200 kilograms of farmed fish per person. More than 300 people are directly employed on Faroese fish farms. In addition, a work force of 1000 is employed in processing and transporting fish, manufacturing fish food, and other related industries. In recent years, aquaculture has become of greater importance to the Faroese economy than in any other country. In 2001-2003, fish farming products constituted approximately 25% of total earnings from exported goods. Wild marine fisheries products constitute the only other major export, accounting for approximately 70% of the earnings from exported goods.

However, aquaculture faces growing problems. Financial strains are increasing due to salmon diseases and the large reduction in market prices. Some untreatable diseases, notably infectious salmon anemia and bacterial kidney disease, occur with unusual frequency in the Faroe Islands. The industry needs an influx of capital if they are to continue the high level of production of recent years and the problems with diseases and low market prices make such an influx unlikely. It has thus been predicted that production during 2004-2006 will drop as shown in the figure. A warming climate can have both positive and negative effects. If warming does not exceed about 5°C, fish growth rates and the length of the growing season are expected to increase. Greater increases in temperature could exceed the thermal tolerance of the fish. Warming also tends to increase incidence of fish diseases and toxic algal blooms.

Possible Changes in Faroe Island Aquaculture Production

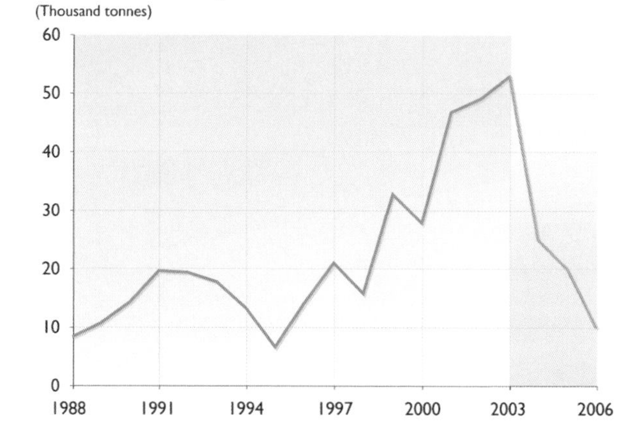

The production of farmed Atlantic salmon and rainbow trout 1988-2003. The red line is a projection for 2004-2006. The projected decrease reflects problems caused by fish diseases and economic issues. Climate change adds additional uncertainties.

4 Animal species' diversity, ranges, and distrubution will change.

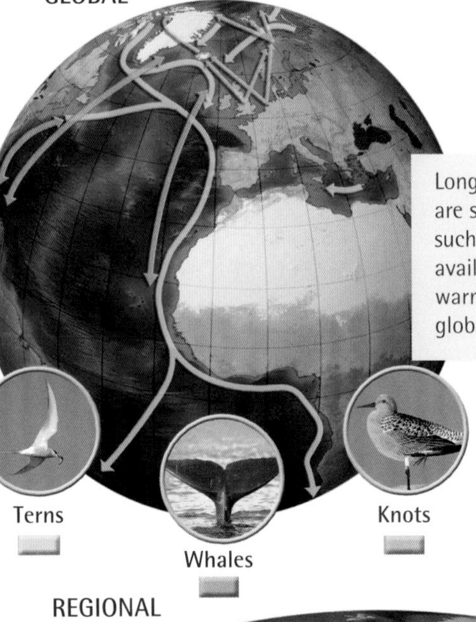

Long-distance animal migration routes are sensitive to climate-related changes such as alterations in habitat and food availability. The amplification of warming in the Arctic thus has global implications for wildlife.

Animal Species on Land

Arctic animals on land include small plant-eaters like ground squirrels, hares, lemmings, and voles; large plant-eaters like moose, caribou/reindeer, and musk ox; and meat-eaters like weasels, wolverine, wolf, fox, bear, and birds of prey.

Climate-related changes are likely to cause cascading impacts involving many species of plants and animals. Compared to ecosystems in warmer regions, arctic systems generally have fewer species filling similar roles. Thus when arctic species are displaced, there can be important implications for species that depend upon them. For example, mosses and lichens are particularly vulnerable to warming. Because these plants form the base of important food chains, providing primary winter food sources for reindeer/caribou and other species, their decline will have far-reaching impacts throughout the ecosystem. A decline in reindeer and caribou populations will affect species that hunt them (including wolves, wolverines, and people) as well as species that scavenge on them (such as arctic foxes and various birds). Because some local communities are particularly dependent on reindeer/caribou, their well-being will also be affected.

REGIONAL LEVEL

Polar Bears
Trees and shrubs
Whales
Birds
Salmon
Reindeer

Ice crust formation resulting from freeze-thaw events affects most arctic land animals by encapsulating their food plants in ice, severely limiting forage availability and sometimes killing the plants. Lemmings, musk ox, and

At the regional level, vegetation and the animals associated with it will shift in response to warming, thawing permafrost, and changes in soil moisture and land use. Range shifts will be limited by geographical barriers such as mountains and bodies of water. Shifts in plankton, fish, and marine mammals and seabirds, particularly those associated with the retreating ice edge, will result from changes in air and ocean temperatures and winds.

LANDSCAPE LEVEL

Retreat of ice edge and its diversity
Extractive industry and pollution
Drying of ponds
Advance of trees, shrubs and southern species
Thawing permafrost
Corridors and fragmentation of habitats
Mountain barrier
N
Forest fire and pests

At the landscape level, shifts in the mosaic of soils and related plant and animal communities will be associated with warming-driven drying of shallow ponds, creation of new wet areas, land use change, habitat fragmentation, and pests and diseases. These changes will affect animals' success in reproduction, dispersal, and survival, leading to losses of northern species and range extensions of southern species.

Science Chapters: Future Climate 4 | Tundra & Polar Deserts 7 | Nature Conservation 10

reindeer/caribou are all affected, and dramatic population crashes resulting from ice crusting due to freeze-thaw events have been reported and their frequency appears to have increased over recent decades. The projected winter temperature increase of over 6°C by late this century (average of the five ACIA model projections) could result in an increase in alternating periods of melting and freezing. Inuit of Nunavut, Canada report that caribou numbers decrease in years when there are many freeze-thaw cycles. Swedish Saami note that over the last decade, autumn snow lies on unfrozen ground rather than on frozen ground in summer grazing areas and this results in rotten and poor quality spring vegetation.

Warming leads to other cascading impacts on arctic land animals. In winter, lemmings and voles live and forage in the space between the frozen ground of the tundra and the snow, almost never appearing on the surface. The snow provides critical insulation. Mild weather and wet snow lead to the collapse of these under-snow spaces, destroying the burrows of voles and lemmings, while ice crust formation reduces the insulating properties of the snow pack vital to their survival. Well-established population cycles of lemmings and voles are no longer seen in some areas. Declines in populations of these animals can lead to declines in the populations of their predators, particularly those predators that specialize in preying on lemmings, such as snowy owls, skuas, weasels, and ermine. A decline in lemming populations would be very likely to result in an even stronger decline in populations of these specialist predators. More generalist predators, such as the arctic fox, switch to other prey species when lemming populations are low. Thus, a decline in lemmings can also indirectly result in a decline in populations of other prey species such as waders and other birds.

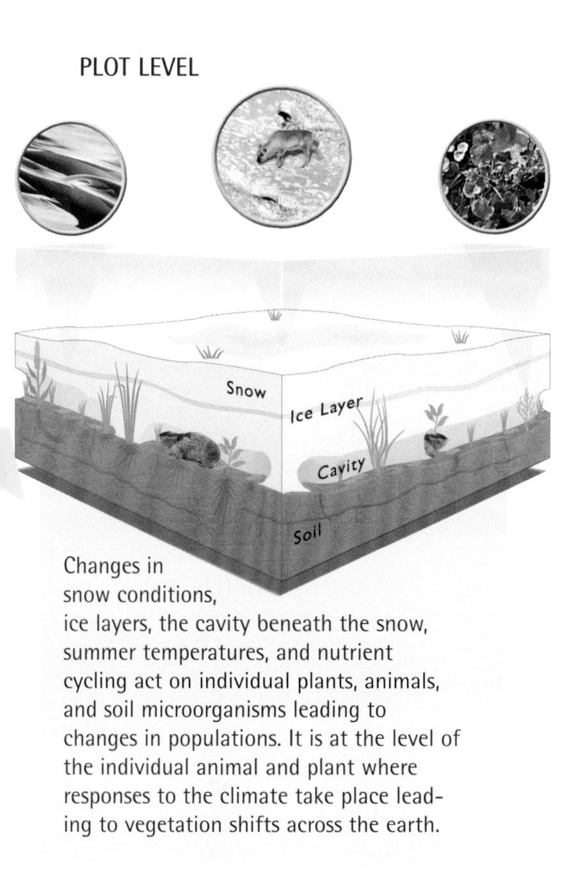

Changes in snow conditions, ice layers, the cavity beneath the snow, summer temperatures, and nutrient cycling act on individual plants, animals, and soil microorganisms leading to changes in populations. It is at the level of the individual animal and plant where responses to the climate take place leading to vegetation shifts across the earth.

Cascading Impacts in a Changing Climate

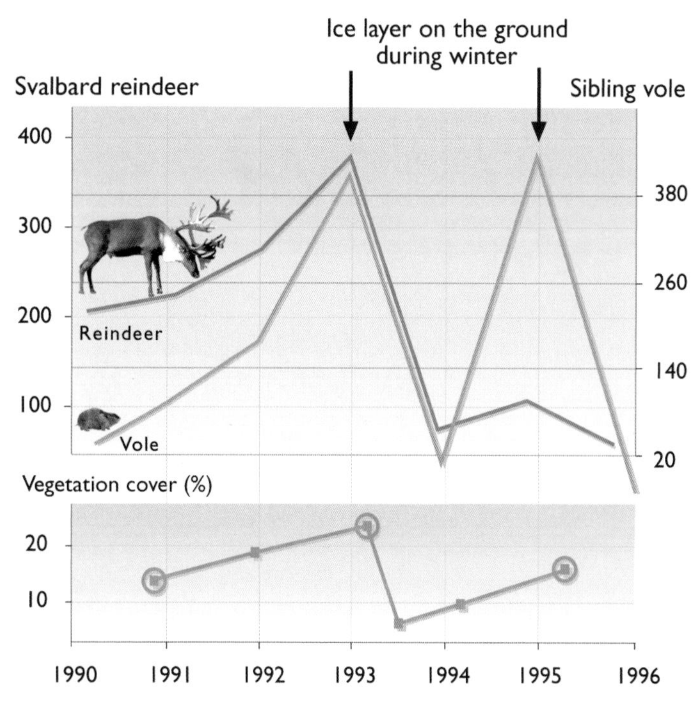

Population dynamics (number of individuals in a particular area) of Svalbard reindeer and sibling voles on Svalbard, along with observed (circles) and projected (squares) changes in vegetation.

4 Animal species' diversity, ranges, and distrubution will change.

"During autumn times, the weather fluctuates so much, there is rain and mild weather. This ruins the lichen access for the reindeer. In some years this has caused massive loss of reindeers. It is very simple - when the bottom layer freezes, reindeer cannot access the lichen. This is extremely different from the previous years. This is one of the reasons why there is less lichen. The reindeer has to claw to force the lichen out and the whole plant comes, complete with roots [bases]. It takes extremely long for a lichen to regenerate when you remove the roots of the lichen."

Heikki Hirvasvuopio
Kakslauttanen, Finland

Caribou/Reindeer

Caribou (North American forms of *Rangifer tarandus*) and reindeer (Eurasian forms of the same species) are of primary importance to people throughout the Arctic for food, shelter, fuel, tools, and other cultural items. Caribou and reindeer herds depend on the availability of abundant tundra vegetation and good foraging conditions, especially during the calving season. Climate-induced changes to arctic tundra are projected to cause vegetation zones to shift significantly northward, reducing the area of tundra and the traditional forage for these herds. Freeze-thaw cycles and freezing rain are also projected to increase. These changes will have significant implications for the ability of caribou and reindeer populations to find food and raise calves. Future climate change could thus mean a potential decline in caribou and reindeer populations, threatening human nutrition for many indigenous households and a whole way of life for some arctic communities.

Peary Caribou

The present reduced state of Peary caribou (a small, white sub-species found only in West Greenland and Canada's arctic islands) is serious enough that a number of communities have limited and even banned their subsistence harvests of the species. The number of Peary caribou on Canada's arctic islands dropped from 26 000 in 1961 to 1000 by 1997, causing the sub-species to be classified as endangered in 1991. The decline of Peary caribou appears to have been caused by autumn rains that iced the winter food supply and crusted the snow cover, limiting access to forage. Also, annual snowfall in the western Canadian Arctic increased during the 1990s and the three heaviest snowfall winters coincided with Peary caribou numbers on Bathurst Island dropping from 3000 to an estimated 75 between 1994 and 1997.

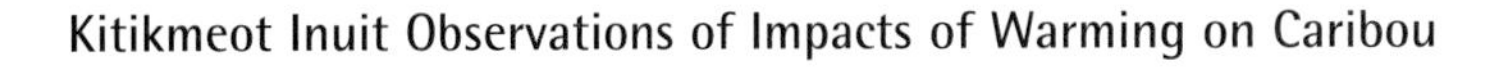

Kitikmeot Inuit Observations of Impacts of Warming on Caribou

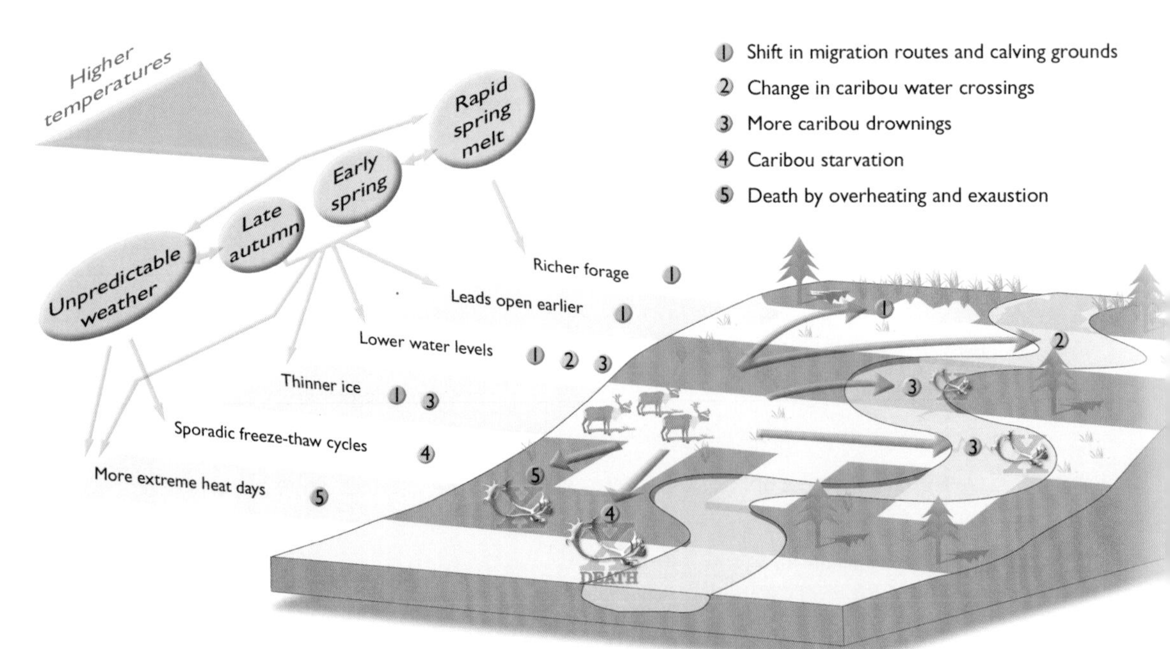

The Porcupine Caribou Herd

The Porcupine Caribou Herd is one of approximately 184 wild herds of caribou globally, the eighth largest herd in North America, and the largest migratory herd of mammals shared between the United States and Canada. The Porcupine Herd has been monitored periodically since the early 1970s. The population grew at about 4% per year from the initial censuses to a high of 178 000 animals in 1989. During the same period, the populations of all major herds increased throughout North America, suggesting that they were responding to continental-scale events, presumably climate-related. Since 1989, the herd has declined at 3.5% per year to a low of 123 000 animals in 2001. The Porcupine Caribou Herd appears to be more sensitive to the effects of climate change than other large herds.

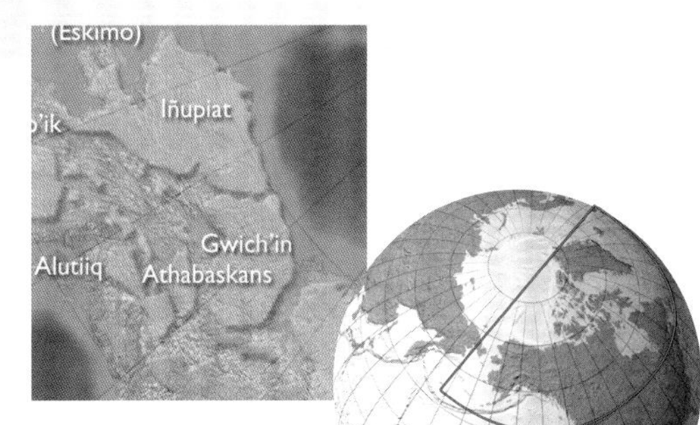

Caribou Ranges and Indigenous Peoples of North America

The ecosystem defined by the range of the Porcupine Herd includes human communities, most of which depend on harvesting caribou for subsistence. Among these are the Gwich'in, Iñupiat, Inuvialuit, Han, and Northern Tuchone whose relationships with this herd have persisted over many millennia. Historically, caribou have served as a critical resource, allowing northern indigenous people to survive the hardships of the severe arctic and sub-arctic conditions. Times of caribou scarcity were often accompanied by great human hardship. Records and oral accounts suggest that periods of caribou scarcity in North America coincided with periods of climatic change.

Today, caribou remain an important component of the mixed subsistence-cash economy, while also enduring as a central feature of the mythology, spirituality, and cultural identity of Indigenous Peoples. The harvesting of the Porcupine Caribou Herd varies from year to year, depending on the distribution of animals, communities' access to them, and community need. The total annual harvest from this herd typically ranges from approximately 3000 to 7000 caribou. Responsibility for management of the herd and protection of its critical habitat is shared in Canada between those who harvest the caribou (mostly Indigenous Peoples) and the government agencies with legal management authority.

Harvest of Porcupine Caribou Herd by User Group

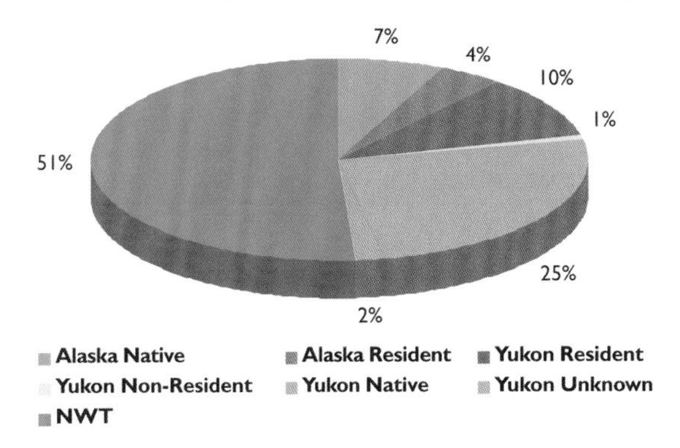

This chart apportions annual average harvest of the Porcupine Caribou Herd in northwestern Canada and northeastern Alaska by user group. Approximately 89% of the harvest is taken in Canada, and more than 90% of the total harvest is taken by indigenous communities.

Key Finding #4

4 Animal species' diversity, ranges, and distrubution will change.

The Gwich'in and the Porcupine Caribou Herd

The Gwich'in have been living in close relationship with the Porcupine Caribou Herd for thousands of years. Gwich'in communities are named for the rivers, lakes, and other aspects of the land with which they are associated. The Vuntut (lake) Gwich'in of Old Crow (population 300) in Canada's Yukon, are located in the center of the Porcupine Caribou Herd's range, providing opportunities to intercept caribou during both their autumn and spring migrations. Average harvest of caribou is as high as five animals per person per year. Sharing among households in the community and with neighboring communities is an important cultural tradition and is also believed to help ensure future hunting success.

"If I were a caribou, I'd be pretty confused right now."

Stephen Mills
Old Crow, Canada

Climate-related factors influence the health of the animals and the herd's seasonal and annual distribution and movement. Climate-related factors also affect hunters' access to hunting grounds, for example, through changes in the timing of freeze-up and break-up of river ice and the depth of snow cover.

Every spring for many generations, the Porcupine Caribou Herd has crossed the frozen Porcupine River to its calving grounds in the Arctic National Wildlife Refuge in Alaska. In recent years, the herd has been delayed on its northern migration as deeper snows and increasing freeze-thaw cycles make their food less accessible, increase feeding and travel time, and generally reduce the health of the herd. At the same time, river ice is thawing earlier in the spring. Now when the herd reaches the river, the river is no longer frozen. Some cows have already calved on the south side and have to cross the rushing water with their newborn calves. Thousands of calves have been washed down the river and died, leaving their mothers to proceed without them to the calving grounds.

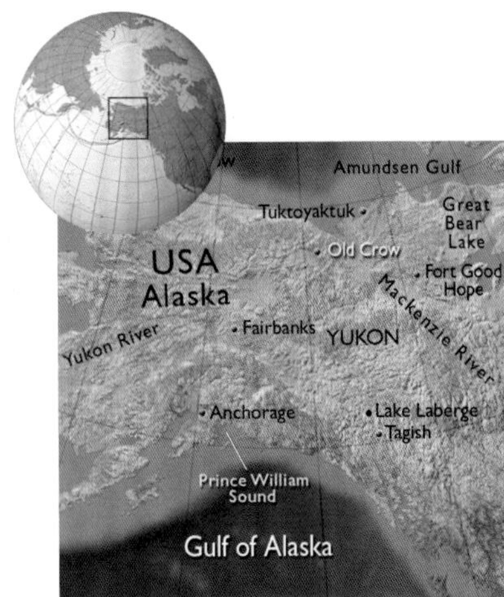

Potential Climate Change Impacts on the Porcupine Caribou Herd

Climate Change Condition	Impact on Habitat	Impact on Movement	Impact on Body Condition	Impact on Productivity	Management Implications
Earlier Snow Melt on Coastal Plain	Higher plant growth rate	Core calving grounds move further north	Cows replenish protein reserves faster	Higher probability of pregnancy	Concern over development on northern portion of present core calving area
		Less use of foothills for calving	Higher calf growth rate		
			Lower predation risk	Higher June calf survival	
Warmer, Drier Summer	Earlier peak biomass	Movement out of Alaska earlier in season	Increased harassment resulting in lower body condition	Lower probability of pregnancy	Protection of insect relief areas important
	Plants harden earlier	More use of coastal zone while in Alaska			
	Reduction in mosquito breeding sites	More dependence on insect relief areas, especially from mid- to late July			
	Significant increase in oestrid activity				
	Greater frequency of fire on winter range				
	Fewer "mushroom" years				
Warmer, Wetter Autumn	More frequent icing conditions	Caribou abandon ranges with severe surface icing	Unknown	Unknown	Protection of low snow regions
Warmer, Wetter Winter	Deeper denser snow	Increased use of low snow regions	Greater over-winter weight loss	Maternal bond broken earlier	
		Later to leave winter range			
Warmer Spring	More freeze / thaw days, snow forms ice layers	Move to windswept slopes	Accelerated weight loss in spring	Higher wolf predation due to use of windswept slopes	Concern over timing and location of spring migration in relation to harvest
	Faster spring melt	Faster spring migration			Lower productivity due to high spring mortality
Overall Effect	Calving range improves, summer, autumn and winter ranges probably lower quality	Seasonal distribution less predictable, timing less predictable	Improved June condition but later summer reduced condition, more rapid weight loss in winter and early spring	High pregnancy rates but overall lower survival and recruitment; Shift mortality later in year (late winter, spring); Herd more likely to decline	Need to assess habitat protection in relation to climate trends
	Extremes (such as very deep snow or very late melt) hard to adapt to				Need to factor climate change impacts on harvest levels
					Need to communicate impacts of climate on harvest patterns and timing
					Need to set up monitoring programs

"Sometimes when they're supposed to show up, they don't show up. Sometimes they show up when they're not supposed to show up... We've got 15 villages in northeast Alaska and north Yukon Territory, and some in Northwest Territory, where the same people are depending on one caribou herd. We're caribou people... and we all depend on that same caribou herd that migrates through our villages."

Sarah James
Arctic Village, Alaska

4 Animal species' diversity, ranges, and distrubution will change.

Freshwater Ecosystems

Freshwater ecosystems in the Arctic include rivers, lakes, ponds, and wetlands, their plant and animal inhabitants, and their surroundings. Animal life in these ecosystems includes fish such as salmon, brown and lake trout, Arctic char, cisco, whitefish, and grayling; mammals such as beavers, otters, mink, and muskrats; waterfowl like ducks and geese; and fish-eating birds such as loons, osprey, and bald eagles.

Climate change will directly and indirectly affect these animals and related biodiversity. Many of the effects will result from climate-induced physical and chemical changes to freshwater habitats. Of particular importance are increasing water temperatures and precipitation, thawing of permafrost, reductions in duration and thickness of lake and river ice, changes in the timing and intensity of runoff, and increased flows of contaminants, nutrients, and sediments. Freshwater ecosystems are also important to marine systems because they act as intermediaries between land and ocean systems, transferring inputs received from the land to the marine environment. Some examples of important physical and chemical changes that will affect freshwater ecosystems include increasing water temperatures, thawing permafrost, ice cover changes on rivers and lakes, and increasing levels of contaminants.

As thawing permafrost and other climate change impacts cause freshwater habitats to disappear, re-form, and be modified, major shifts in species and their use of aquatic habitats are likely.

Water Temperature Increases

Increases in water temperature are likely to make it impossible for some species to remain in parts of streams and lakes they formerly inhabited. Less than optimum thermal conditions, combined with other possible effects, such as competition from invasive species moving in from the south, may significantly shrink the ranges of some arctic freshwater species, such as the broad whitefish, Arctic char, and Arctic cisco.

Permafrost Thawing

As rising temperatures thaw frozen soils, drainage of water from lakes into groundwater can occur, eventually eliminating the aquatic habitat in the area. On the other hand, collapsing of the ground surface due to permafrost thawing can create depressions where new wetlands and ponds can form, adding to aquatic habitat in these locations. The balance of these changes is not known, but as freshwater habitats disappear, re-form, and are modified, major shifts in species and their use of aquatic habitats are likely.

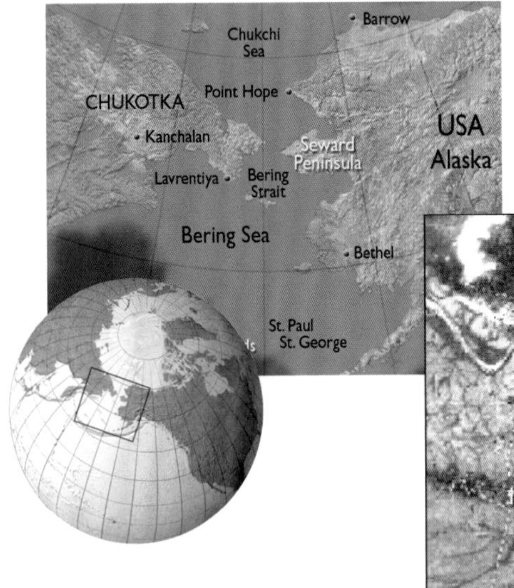

Shrinking Tundra Ponds

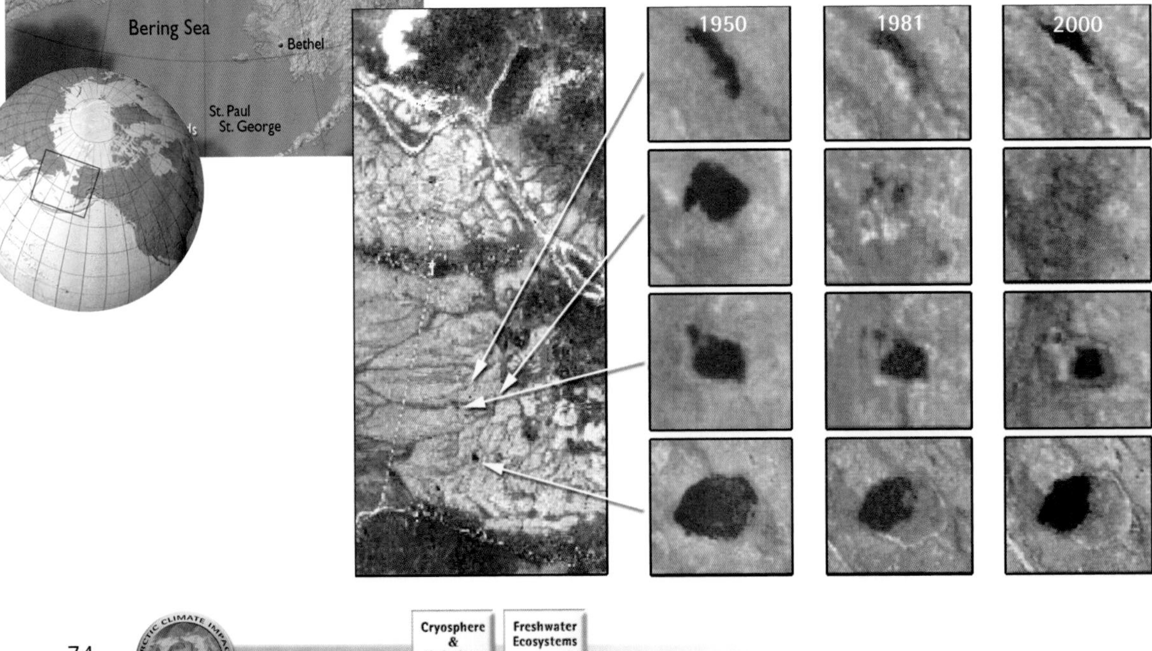

The ponds pictured to the left are on Alaska's Seward Peninsula. Of 24 ponds studied in this region, 22 decreased in area between 1951 and 2000. Numerous tundra ponds have decreased in surface area over the last 50 years. A probable mechanism for this shrinkage is internal drainage as shallow permafrost thaws.

Science Chapters: Cryosphere & Hydrology 6 Freshwater Ecosystems 8

Ice Cover Changes on Rivers and Lakes

Lake and river ecology are strongly affected by ice cover and the timing of the spring melt. Ecological impacts will result from changes in the timing of ice break-up, which strongly affects supplies of nutrients, sediments, and water that are essential to the health of delta and floodplain ecosystems. Changes in ice timing and types also affect water temperature and levels of dissolved oxygen (required by most living things in the system). Changes in species composition and diversity and food web structure are among the expected results of these climate-induced changes. Reduced ice cover will also dramatically increase the exposure and related damage of underwater life forms to ultraviolet radiation.

Later freeze-up and earlier break-up of river and lake ice have combined to reduce the ice season by one to three weeks, depending on location, over the past 100 years. This trend is strongest over the western parts of Eurasia and North America and is projected to continue over the next 100 years, causing a general reduction in ice cover on arctic rivers and lakes, with the greatest reductions projected in the northernmost lands. Freeze-up and break-up dates respond strongly to warming because as ice melts, it results in further warming of the surface, causing more melting, more warming, and so on. Longer ice-free periods are projected to increase evaporation, leading to lower water levels, though this may be countered by the increase in precipitation projected to result from the greater availability of ocean moisture (where sea ice has retreated). These changes will affect whether the northern peatlands will absorb or release the greenhouse gases carbon dioxide and methane. Low flow and flood patterns will change, as will levels of sediments carried by rivers to the Arctic Ocean.

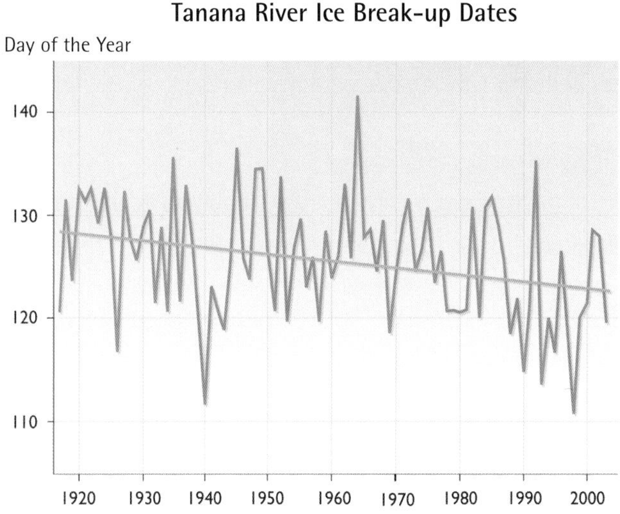

The graph shows the ice break-up dates for the Tanana River at Nenana, Alaska over the last 80 years. Though there is considerable variability from year to year, there is a trend toward earlier break-up by over a week.

Contaminants

Warming is very likely to accelerate rates of contaminant transfer to the Arctic and increased precipitation is very likely to increase the amount of persistent organic pollutants and mercury that are deposited in the region. As temperatures rise, snow and ice accumulated over years to decades will melt, and the contaminants stored within will be released in melt water. Permafrost thawing may similarly mobilize contaminants. This will increase episodes of high contaminant levels in rivers and ponds that may have toxic effects on aquatic plants and animals and also increase transfer of pollutants to marine areas. These impacts will be amplified by lower water levels as higher temperatures increase evaporation (possibly countered by increasing precipitation in some areas). Increasing contaminant levels in arctic lakes will accumulate in fish and other animals, becoming magnified as they are transferred up the food chain.

Freshwater Food Web

Predatory birds: Gulls Peregrine falcon
Herbivorous birds: Ducks Geese
Predatory mammals: Arctic Fox
Insect larvae
Birds: Cranes Plovers
Carnivorous zooplankton
Herbivorous zooplankton
Pelagic and benthic algae
Vegetation: Sedges Grasses Duckweed
Second level predators ← First level predators ← Grazers and filter feeders ← Primary producers

4 Animal species' diversity, ranges, and distrubution will change.

Freshwater Fish

Southernmost species are projected to shift northward, competing with northern species for resources. The broad whitefish, Arctic char, and Arctic cisco are particularly vulnerable to displacement as they are wholly or mostly northern in their distribution. As water temperatures rise, spawning grounds for cold-water species will shift northward and are likely to be diminished. As southerly fish species move northward, they may introduce new parasites and diseases to which arctic fish are not adapted, increasing the risk of death for arctic species. The implications of these changes for both commercial and subsistence fishing in far northern areas are potentially devastating as the most vulnerable species are often the only fishable species present. In some southern mainland areas of the Arctic, new arrivals from the south may also bring new opportunities for fisheries, and increased productivity of some northern fish populations due to higher growth may allow for increased fishing of some species.

Arctic Char

The Arctic char is the northernmost freshwater fish in the world and occurs throughout the Arctic. Some populations are locked in lakes where they feed on midge larvae and grow very slowly. Other populations migrate to the sea in summer where they feed on crustaceans and small fish, and char in these populations grow more quickly. Increasing water temperatures in freshwaters, estuaries, and marine near-shore areas are likely to increase growth of both types of char, especially in the mid-latitudes of their distribution, assuming that there is also a parallel general increase in food chain productivity. This is likely to increase fishing opportunities, but may be offset by the effects of competition from new fish species.

The implications of these changes for both commercial and subsistence fishing in far northern areas are potentially devastating as the most vulnerable species are often the only fishable species present.

Research on Arctic char in Resolute Lake, Canada suggests that rising temperatures cause an increase in respiration, which increases the accumulation of heavy metals in the fish. In addition, other climate-related changes described on the previous page are expected to increase the levels of contaminants in lakes. Furthermore, reduced ice cover in lakes, increased mixing between water layers, and other warming-induced changes are projected to result in lakes retaining more of the contaminants that flow into them.

Arctic Grayling

The Arctic grayling is a stream fish with about a 12-year lifespan. In some northern locations, it is the only species of fish that occupies local streams. In Toolik Lake (a small lake on Alaska's tundra), 25 years of data have been collected on

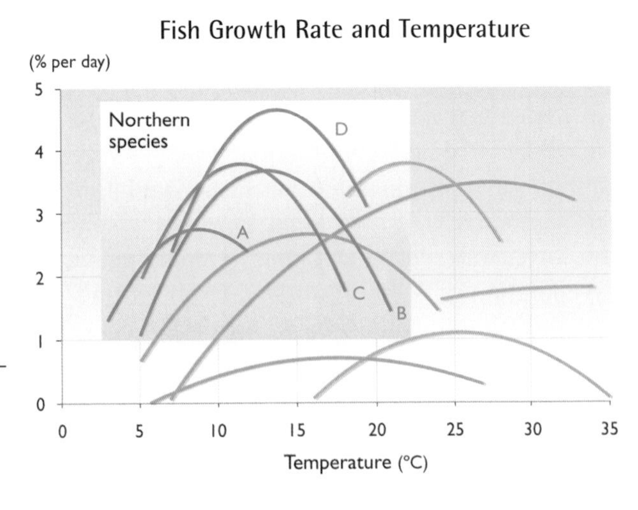

These growth curves (in percent per day) for various fish species illustrate that growth typically increases with rising temperature up to a certain point and then declines as temperature continues to rise. Northern species (A. Arctic char, B. lake cisco, C. lake trout, and D. brook trout - all in blue) are grouped toward the lower temperatures on the left, and have a more peaked curve, indicating only narrow and typically low temperature ranges over which optimal growth is achieved. This suggests that their ability to adapt to a warming climate is likely to be quite limited. The unlabeled growth curves are for various lower latitude species.

the grayling, tracking each individual fish in the stream. Results indicate that while young grayling do well in warmer water, adults fare poorly, actually losing weight in warm years. Projected climate warming is thus likely to cause the elimination of this population, with no opportunities for other species to naturally come into the lake.

Lake Trout

Long-term studies project that a warmer future will severely stress lake trout, with related impacts on the food web. Impacts on lake trout will be most severe in smaller lakes in the southern part of the trout's range in the Arctic; effects in larger, more northern lakes may be positive, at least in the short-term. Long-term studies in Toolik Lake, Alaska project that a warmer future is likely to result in the elimination of this lake trout population. Research suggests that a 3˚C rise in July surface water temperature could cause first-year lake trout to need to consume eight times more food than is currently necessary just to maintain adequate body condition. This requirement greatly exceeds current food availability in the lake.

Furthermore, the projected future combination of higher temperatures, a longer open-water season, and increased phosphorus in the water (released into streams as permafrost thaws) is expected to increase production of small aquatic life forms that consume oxygen, thereby reducing oxygen concentrations in deeper water to a level below that needed by lake trout (and some other living things), thus reducing the bottom-water habitat. With surface water warming beyond the threshold required for these fish, the trout will be squeezed into a shrinking habitat between the inhospitable conditions near the surface and those at the lake bottom. The loss of the lake trout, the top predator in this system, is likely to have cascading impacts through the food web, with major impacts on both the structure and functioning of the ecosystem.

"New species of plants have arrived. We never saw them before. This is what we have observed. New plants have arrived here and on tundra. Even there are arrival species in the river, previously known in middle parts of Russia. This past summer and the previous were very hot here. Rivers and lakes are filled with small-flowered kind of duckweed, and the lake started to bloom. Life of the fish is more difficult, and likewise people's fishing opportunities, as lakes grow closed up with the new plants."

Larisa Avdeyeva
Lovozero, Russia

Aquatic Mammals and Waterfowl

The distributions of aquatic mammals and waterfowl are likely to expand northward as habitats change with warming. Seasonal migration is also likely to occur earlier in spring and later in autumn if temperatures are warm enough. Breeding ground suitability and access to food will be the primary drivers of changes in migration patterns. For example, wetlands are important breeding and feeding grounds for ducks and geese in spring. As permafrost thaws, more wetlands (formed when previously frozen ground collapses) are likely to appear, promoting the earlier northward migration of southerly wetland species or increasing the abundance and diversity of current high latitude species. However, a parallel earlier timing of the availability of local food must also occur for these outcomes to be realized.

Mammal and bird species moving northward are likely to carry new diseases and parasites that pose new threats to arctic species. Another potential threat from the northward movement of southerly species is that they may out-compete northern species for habitat and resources. Northerly species may have diminished reproductive success as suitable habitat either shifts northward or declines in availability or access.

5 Many coastal communities and facilities face increasing exposure to storms.

Rising temperatures are altering the arctic coastline and much larger changes are projected to occur during this century as a result of reduced sea ice, thawing permafrost, and sea-level rise. Thinner, less extensive sea ice creates more open water, allowing stronger wave generation by winds, thus increasing wave-induced erosion along arctic shores. Sea-level rise and thawing of coastal permafrost exacerbate this problem. In some areas, an eroding shoreline combines coarse sediments with frozen seawater, creating huge blocks of ice that carry sediments for distances of over 100 kilometers. These sediment-laden ice blocks pose dangers to ships and further erode the coastline as they are carried along by the winds. Higher waves will create even greater potential for this kind of erosion damage.

"Some of our communities are eroding into the ocean in front of our eyes because of the decrease in the multi-layered ice, which is allowing for larger storms to roll in."

Duane Smith
Inuit Circumpolar Conference, Canada

Arctic Coastal Areas Susceptible to Erosion

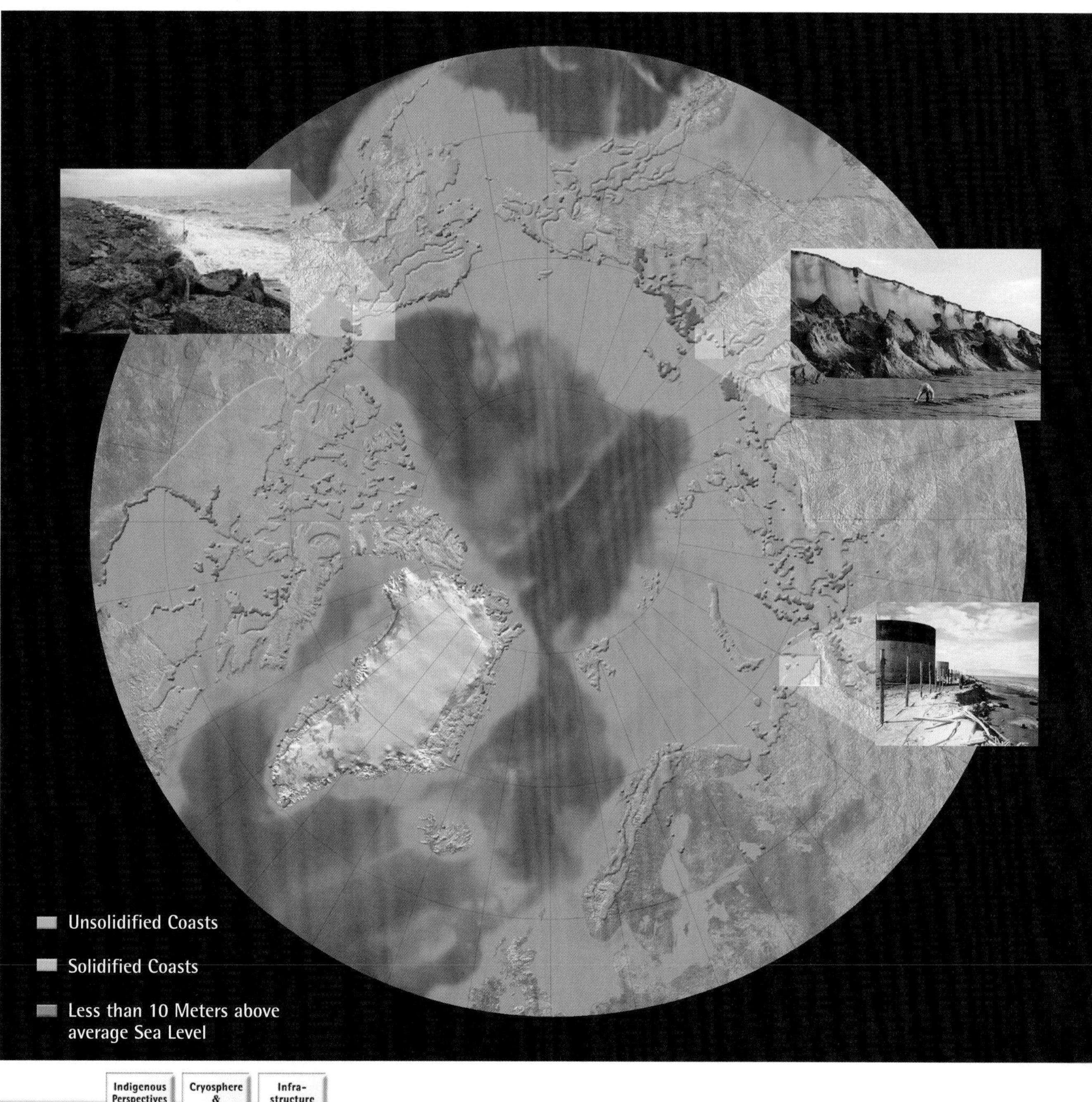

Rising sea level is very likely to inundate marshes and coastal plains, accelerate beach erosion, exacerbate coastal flooding, and force salt water into bays, rivers, and groundwater, not only in the Arctic, but also around the world. Local sea-level rise depends on how much the oceans are expanding as well as on whether the local coastline is rising or subsiding due to forces affecting the earth's crust (such as rebound from the last ice age). Arctic coasts show a wide variation in these trends, although low-lying coastal plains of the Arctic are generally not rising, making them more vulnerable to adverse effects of sea-level rise. Higher sea level at the mouths of rivers and bays will allow salt water to penetrate further inland. Storms that bring more intense rainfall at the coast will increase erosion by runoff and the amount of mobile sediment in coastal waters.

Coastal erosion will pose increasing problems for some ports, tanker terminals, and other industrial facilities, as well as for coastal villages.

Coastal regions with underlying permafrost are especially vulnerable to erosion as ice beneath the seabed and shoreline thaws from contact with warmer air and water. Though little specific monitoring has yet been done, generally, the projected increase in air and water temperature, reduction in sea ice, and increase in height and frequency of storm surges are expected to have a destabilizing effect on coastal permafrost, resulting in increased erosion. Low-lying ice-rich permafrost coasts are thus most vulnerable to wave-induced erosion. One result of this erosion is that more sediment will be brought to coastal waters, adversely affecting marine ecosystems. Increased coastal permafrost degradation could also result in greater releases of carbon dioxide and methane. Coastal erosion will pose increasing problems for some ports, tanker terminals, and other industrial facilities, as well as for coastal villages. Some towns and industrial facilities are already suffering severe damage and some are facing relocation as warming begins to take its toll on arctic coastlines.

In the Alaskan village of Nelson Lagoon, residents have built increasingly strong break walls along the shore, only to see them destroyed by increasingly violent coastal storms. Their break walls were designed to brace the shore ice, which would in turn provide the major buffer from winter storm wave action. As winters have warmed, the buffer provided by the shore ice has been lost, allowing the full force of the waves to surge against the wall and the village. The pipeline that provides drinking water for the village was also threatened when storm waves eroded soil cover and caused a breach in the line.

The vulnerability of a coastline to erosion depends on sea level, the properties of the coastal materials, and environmental factors such as tectonic forces and wave action. Unsolidified arctic coasts (in green) containing variable amounts of ground ice, are more susceptible to erosion than solidified coasts (in orange). Unstable coastal environments are shown in the inset photographs from the Pechora, Laptev, and Beaufort Sea coasts. Tectonic forces create uplift in some places, including the Canadian Archipelago, Greenland, and Norway, and subsidence in others, such as along the Beaufort Sea and Siberian coasts. Areas (in red) in which elevation is less than 10 meters above average sea level are particularly vulnerable.

5 Many coastal communities and facilities face increasing exposure to storms.

Shishmaref, Alaska Faces Evacuation

The village of Shishmaref, located on an island just off the coast of northern Alaska and inhabited for 4000 years, is now facing the prospect of evacuation. Rising temperatures are causing a reduction in sea ice and thawing of permafrost along the coast. Reduced sea ice allows higher storm surges to reach the shore and the thawing permafrost makes the shoreline more vulnerable to erosion, undermining the town's homes, water system, and other infrastructure.

The problem of coastal erosion has become increasingly serious in Shishmaref in recent years. Over a dozen houses have already had to be moved further from the sea. The 600 residents have watched as one end of their village has been eaten away, losing as much as 15 meters of land overnight in a single storm. The absence of sea ice also deprives the residents of their means of traveling to the mainland to hunt moose and caribou, as they would normally do by early November. Nowadays, the inlet is open water in the autumn.

"I went to school on the mainland, and when I came back, my house was gone. They moved it to the other side of the village, or it would've fallen in."

Leona Goodhope
Shishmaref, Alaska

Village elder, Clifford Weyiouanna says, "The currents have changed, ice conditions have changed, and the freeze-up of the Chukchi Sea has really changed, too. Where we used to freeze up in the last part of October, now we don't freeze 'til around Christmas time. Under normal conditions, the sea ice out there should be four feet [1.2 meters] thick. I went out, and the ice was only one foot [0.3 meter] thick."

Over the last 40 years, villagers estimate that they have lost hundreds of square meters of land. Robert Iyatunguk, erosion coordinator for the village, explains that the retreat of the sea ice is leaving the village more vulnerable to increasingly violent weather. "The storms are getting more frequent, the winds are getting stronger, the water is getting higher, and it's noticeable to everyone in town. If we get 12-14 foot [~4-meter] waves, this place is going to get wiped out in a matter of hours. We're in panic mode because of how much ground we're losing. If our airport gets flooded out, there goes our evacuation by plane."

Severe Erosion in Tuktoyaktuk, Canada

Tuktoyaktuk is the major port in the western Canadian Arctic and the only permanent settlement on the low-lying Beaufort Sea coast. Tuktoyaktuk's location makes it highly vulnerable to increased coastal erosion due to decreased extent and duration of sea ice, accelerated thawing of permafrost, and sea-level rise. The Tuktoyaktuk Peninsula is characterized by sandy spits, barrier islands, and a series of lakes created as thawing permafrost caused the ground the collapse ("thermokarst" lakes). Erosion is already a serious problem in and around Tuktoyaktuk, threatening cultural and archeological sites and causing the abandonment of an elementary school, housing, and other buildings. Successive shoreline protection structures have been rapidly destroyed by storm surges and accompanying waves.

As warming proceeds and sea-level rise accelerates, impacts are expected to include further landward retreat of the coast, erosion of islands, more frequent flooding of low-lying areas, and breaching of freshwater thermokarst lakes and their consequent conversion into brackish or saline lagoons. The current high rates of cliff erosion are projected to increase due to higher sea levels, increased thawing of permafrost, and the increased potential for severe coastal storms during the extended open-water season. Attempts to control erosion at Tuktoyaktuk will become increasingly expensive as the surrounding coastline continues to retreat. The site could ultimately become uninhabitable.

Erosion Threatens Russian Oil Storage Facility

The oil storage facility at Varandei on the Pechora Sea was built on a barrier island. Damage to the dunes and beach due to the facility's construction and use have accelerated natural rates of coastal erosion. The Pechora Sea coasts are thought to be relatively stable, except where disturbed by human activity. Because this site has been perturbed, it is more vulnerable to damage due to storm surges and the accompanying waves that will become a growing problem as climate continues to warm. As with the other sites discussed here, the reduction in sea ice, thawing coastal permafrost, and rising sea level are projected to exacerbate the existing erosion problem. This provides an example of the potential for combined impacts of climate change and other human-caused disturbances. Sites already threatened due to human activity are often more vulnerable to the impacts of climate change.

Novaya Zemlya

Kara Sea

Varandei

Amderma

Nelim Nos

Pechora

Ob

6 Reduced sea ice is very likely to increase marine transport and access to resources.

Observations over the past 50 years show a decline in arctic sea-ice extent in all seasons, with the most prominent retreat in summer. Recent studies estimate arctic-wide reductions in annual average sea-ice extent of about 5-10% and a reduction in average thickness of about 10-15% over the past few decades. Measurements taken by submarine sonar in the central Arctic Ocean revealed a 40% reduction in ice thickness in that area. Taken together, these trends indicate an Arctic Ocean with longer seasons of less sea-ice cover of reduced thickness, implying improved ship accessibility around the margins of the Arctic Basin (although this will not be uniformly distributed).

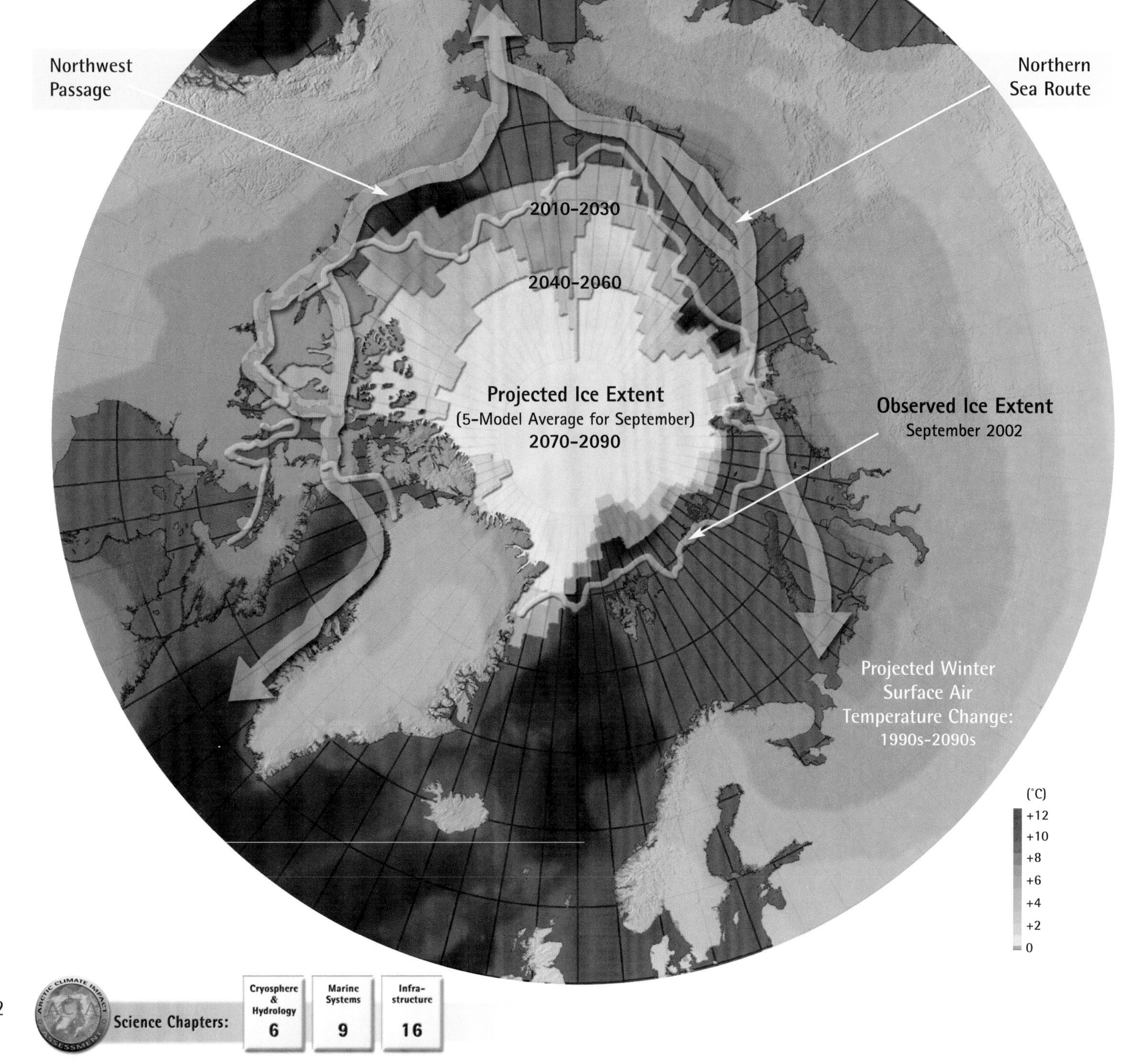

Climate models project an acceleration of this trend, with periods of extensive melting spreading progressively further into spring and autumn. Model projections suggest that sea ice in summer will retreat further and further away from most arctic landmasses, opening new shipping routes and extending the period during which shipping is feasible.

The navigation season is often defined as the number of days per year in which there are navigable conditions, generally meaning less than 50% sea ice concentration. The navigation season for the Northern Sea Route is projected to increase from the current 20-30 days per year to 90-100 days by 2080. Passage is feasible for ships with ice-breaking capability in seas with up to 75% sea-ice concentration, suggesting a navigation season of approximately 150 days a year for these vessels by 2080. Opening of shipping routes and extending the navigation season could have major implications for transportation as well as for access to natural resources.

The Northern Sea Route

The Northern Sea Route (NSR) is the formal Russian name for the seasonally ice-covered marine shipping routes across the north of Eurasia from Novaya Zemlya in the west to the Bering Strait in the east. The NSR is administered by the Russian Ministry of Transport and has been open to marine traffic of all nations since 1991. For trans-Arctic voyages, the NSR represents up to a 40% savings in distance from northern Europe to northeastern Asia and the northwest coast of North America compared to southerly routes via the Suez or Panama Canals.

The NSR also provides regional marine access to the Russian Arctic for ships sailing north from Europe and eastward into the Kara Sea and returning westward to Europe or North America. Regional access from the Pacific side of the NSR is achieved when ships sail through the Bering Strait to ports in the Laptev and East Siberian Seas and return eastward to Asia with cargo. Since 1979, year-round navigation has been maintained by Russian icebreakers in the western region of the NSR, providing a route through Kara Gate and across the Kara Sea to the Yenisey River.

The Russian Arctic holds significant reserves of oil, natural gas, timber, copper, nickel, and other resources that may best be exported by sea. Regional as well as trans-Arctic shipping along the NSR is very likely to benefit from a continuing reduction in sea ice and lengthening navigation seasons.

The satellite image of sea-ice extent for September 16, 2002 provides a good illustration of marine access around the Arctic Basin. Such low summer minimum ice extents create large areas of open water along much of the length of the NSR. The further north the ice edge retreats, the further north ships can sail in open water on trans-Arctic voyages, thereby avoiding the shallow shelf waters and narrow straits of the Russian Arctic.

Northern Sea Route Navigation Season
Projection for 2000-2100

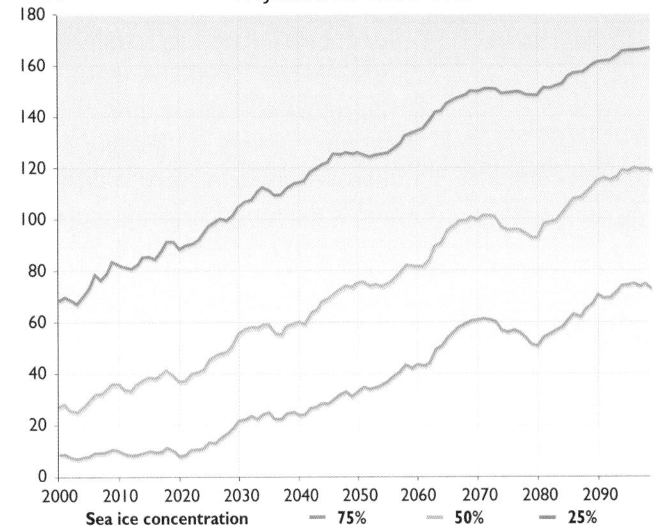

The graph shows the projected increase in days of the navigation season through the Northern Sea Route as an average of five ACIA model projections.

Observed Sea Ice Cover
September 16, 2002

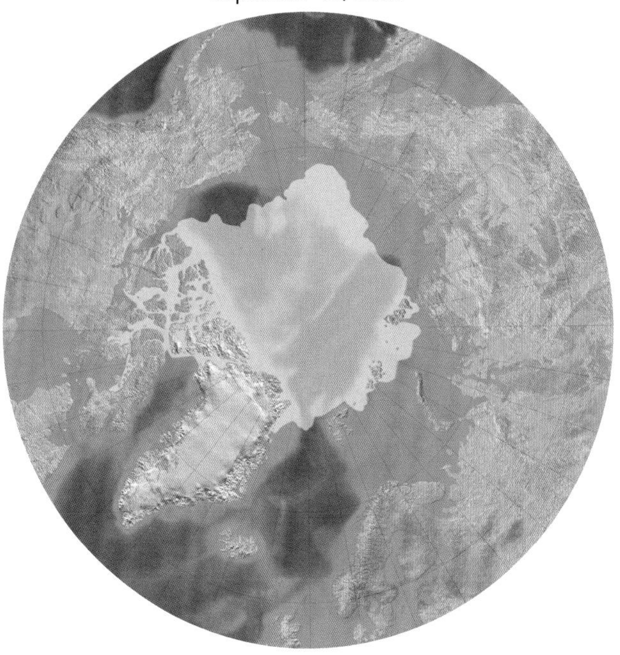

Key Finding #6

6 Reduced sea ice is very likely to increase marine transport and access to resources.

New and revised national and international regulations focusing on marine safety and environmental protection will be necessary.

Sovereignty, Security, and Safety

As the decline in arctic sea ice opens historically closed passages, questions are likely to arise regarding sovereignty over shipping routes and seabed resources. Issues of security and safety could also arise. One impact of the projected increase in marine access for transport and offshore development will be requirements for new and revised national and international regulations focusing on marine safety and environmental protection. Another probable outcome of this growing access will be an increase in potential conflicts among competing users of arctic waterways and coastal seas, for example, in the Northern Sea Route and Northwest Passage. Commercial fishing, sealing, hunting of marine wildlife by indigenous people, tourism, and shipping all compete for use of the narrow straits of these waterways, which are also the preferred routes for marine mammal migration.

With increased marine access in arctic coastal seas – for shipping, offshore development, fishing, and other uses – national and regional governments will be called upon for increased services such as icebreaking assistance, improved ice charting and forecasting, enhanced emergency response in dangerous situations, and greatly improved oil-ice cleanup capabilities. The sea ice, while thinning and decreasing in extent, is likely to become more mobile and dynamic in many coastal regions where fast ice and relatively stable conditions previously existed. Competing marine uses in newly open or partially ice-covered areas will call for increased enforcement presence and regulatory oversight.

Increasing access in the Arctic Ocean will require ships transiting the region to be built to higher construction standards compared with ships operating in the open ocean. International and domestic regulations, designed to enhance maritime safety and marine environmental protection in arctic waters, will need to take into account that each ship will have a high probability of operating in ice somewhere during a voyage. Such ships will have higher construction, operational, and maintenance costs.

Sea Ice Changes Could Make Shipping More Challenging

Not all agree that reduced sea ice, at least in the early part of the 21st century, will necessarily be the boon to shipping that is widely assumed. Recent sea ice changes could, in fact, make the Northwest Passage less predictable for shipping. Studies by the Canadian Ice Service indicate that sea ice conditions in the Canadian Arctic during the past three decades have been characterized by high year-to-year variability; this variability has existed despite the fact that since 1968-1969 the entire region has experienced an overall decrease in sea-ice extent during September. For example, in the eastern Canadian Arctic, some years – 1972, 1978, 1993, and 1996 – have had twice the area of sea ice compared with the first or second year that follows. This significant year-to-year variability in sea ice conditions makes planning for regular marine transportation along the Northwest Passage very difficult.

In addition, results of research at Canada's Institute of Ocean Sciences suggest that the amount of multi-year sea ice moving into the Northwest Passage is controlled by blockages or "ice bridges" in the northern channels and straits of the Canadian Arctic

Archipelago. With a warmer arctic climate leading to higher temperatures and a longer melt season, these bridges are likely to be more easily weakened (and likely to be maintained for a shorter period of time each winter) and the flushing or movement of ice through the channels and straits could become more frequent. More multi-year ice and potentially many more icebergs could thus move into the marine routes of the Northwest Passage, presenting additional hazards to navigation. Thus, despite widespread retreat of sea ice around the Arctic Basin, it is clear that the unusual geography of the Canadian Arctic Archipelago creates exceptionally complex sea ice conditions and a high degree of variability for the decades ahead.

Oil Spills: An Example of the Risks That Accompany Increasing Access

Along with increasing access to shipping routes and resources comes an increasing risk of environmental degradation caused by these activities. One obvious concern involves oil spills and other industrial accidents. A recent study suggests that the effects of oil spills in a high-latitude, cold ocean environment last much longer and are far worse than first suspected.

In 1989, the Exxon Valdez oil tanker slammed into a reef while maneuvering to avoid ice in the shipping lanes and poured 42 million liters (11 million gallons) of crude oil into Alaska's Prince William Sound. The spill was the worst tanker disaster ever in U.S. waters, killing at least 250 000 seabirds and thousands of marine mammals. It forced the closure of commercial fishing grounds and areas traditionally used to gather wild foods. Scientists knew the immediate effects would be devastating but some predicted the environment would recover as soon as the oil weathered and dissipated. Instead, they found that marine life suffered for many years, and continues to suffer, because even tiny patches of remnant oil reduced survival, slowed reproduction, and stunted growth. Lingering oil has created cascading problems for fish, seabirds, and marine mammals.

The recent study found that Valdez oil was still embedded in Prince William Sound beaches in the summer of 2003. "The oil is oozing into holes," said Stanley Rice of the National Marine Fisheries Service laboratory in Juneau, Alaska, who led a team that dug about 1000 pits in beaches in 2003. "There, the oil is like it was two or three weeks after the spill." Sea otters and other animals digging for food are exposed to the oil and its ill effects, Rice said. Studies of sea otters, harlequin ducks, salmon, and shellfish suggest that patches of oil that persist on some beaches release enough hydrocarbons to cause chronic problems that will continue for some species for many years.

Experts say that the overall strategy for arctic spills must be preventative. New regulations for ships, offshore structures, port facilities, and other coastal activities must be designed to reduce the risk of spills through enhanced construction standards and operating procedures. Nevertheless, spills are expected, and spill response operations in the Arctic will be more complex and demanding in ice-covered waters than in Prince William Sound or open seas, especially since effective response strategies have yet to be developed.

7 Thawing ground will disrupt transportation, buildings, and other infrastructure.

Transport on Land

Unlike most parts of the world, arctic land is generally more accessible in winter, when the tundra is frozen and ice roads and bridges are available. In summer, when the top layer of permafrost thaws and the terrain is boggy, travel over land can be difficult. Many industrial activities depend on frozen ground surfaces and many northern communities rely on ice roads for the transport of groceries and other materials. Rising temperatures are already leading to a shortening of the season during which ice roads can be used and are creating increasing challenges on many routes. These problems are projected to increase as temperatures continue to rise. Frost heave and thaw-induced weakening are major factors affecting roadway performance; transportation routes are likely to be particularly susceptible to these effects under changing climatic conditions. In addition, the incidence of mud and rockslides and avalanches are sensitive to the kinds of changes in weather (such as an increase in heavy precipitation events) that are projected to accompany warming.

A Shrinking Ice Road Season

In January 2003, winter road construction in Canada's Northwest Territories was badly behind schedule. Les Shaw, superintendent of transportation for the Fort Simpson region, said the warm weather and lack of snow put winter road and ice bridge construction behind by several weeks. The ice bridge across the Mackenzie River at Fort Providence provided a good example. *"These last two years we've had the ice form across the channel between Christmas and New Year's, which is really strange. Normally it would be doing that at the beginning of December,"* Shaw said. This causes major problems for the territory's mining and oil and gas industries, which rely on the frozen roadways to truck in hundreds of tonnes of supplies for the year.

Impacts of Thawing on the Oil, Gas, and Forestry Industries

Because of warming, the number of days per year in which travel on the tundra is allowed under Alaska Department of Natural Resources standards has dropped from over 200 to about 100 in the past 30 years, resulting in a 50% reduction in days that oil and gas exploration and extraction equipment can be used. These standards, designed to protect the fragile tundra from damage, are currently under review and may be relaxed, raising concerns about potential damage to the tundra. Forestry is another industry that requires frozen ground and rivers. Higher temperatures mean thinner ice on rivers and a longer period during which the ground is thawed. This leads to a shortened period during which timber can be moved from forests to sawmills, and increasing problems associated with transporting wood.

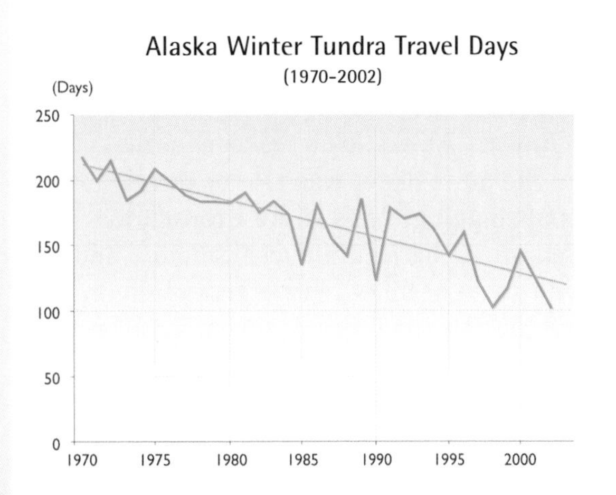

The number of days in which oil exploration activities on the tundra are allowed under Alaska Department of Natural Resources standards has been halved over the past 30 years due to climate warming. The standards are based on tundra hardness and snow conditions and are designed to protect the tundra from damage.

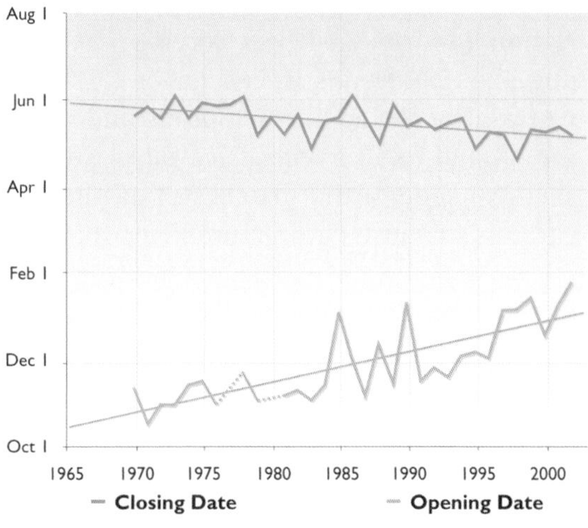

The number of travel days for oil exploration on the Alaskan tundra has been decreasing over recent decades as the opening dates come later and the closing dates come earlier.

Degrading Permafrost

Air temperature, snow cover, and vegetation, all of which are affected by climate change, affect the temperature of the frozen ground and the depth of seasonal thawing. Permafrost temperatures over most of the sub-arctic land areas have increased by several tenths of a °C up to 2°C during the past few decades, and the depth of the active layer is increasing in many areas. Over the next 100 years, these changes are projected to continue and their rate to increase, with permafrost degradation projected to occur over 10-20% of the present permafrost area, and the southern limit of permafrost projected to shift northward by several hundred kilometers.

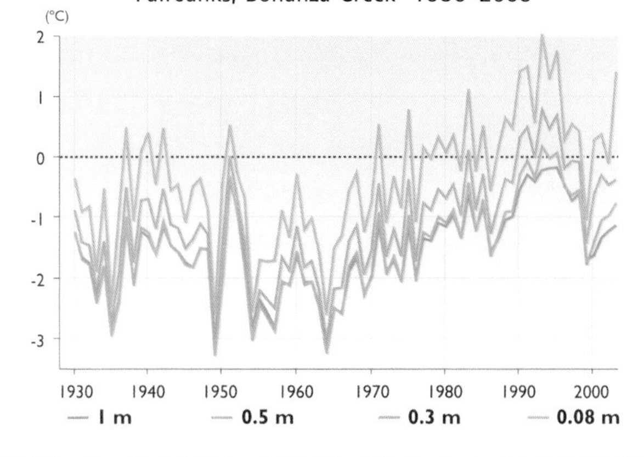

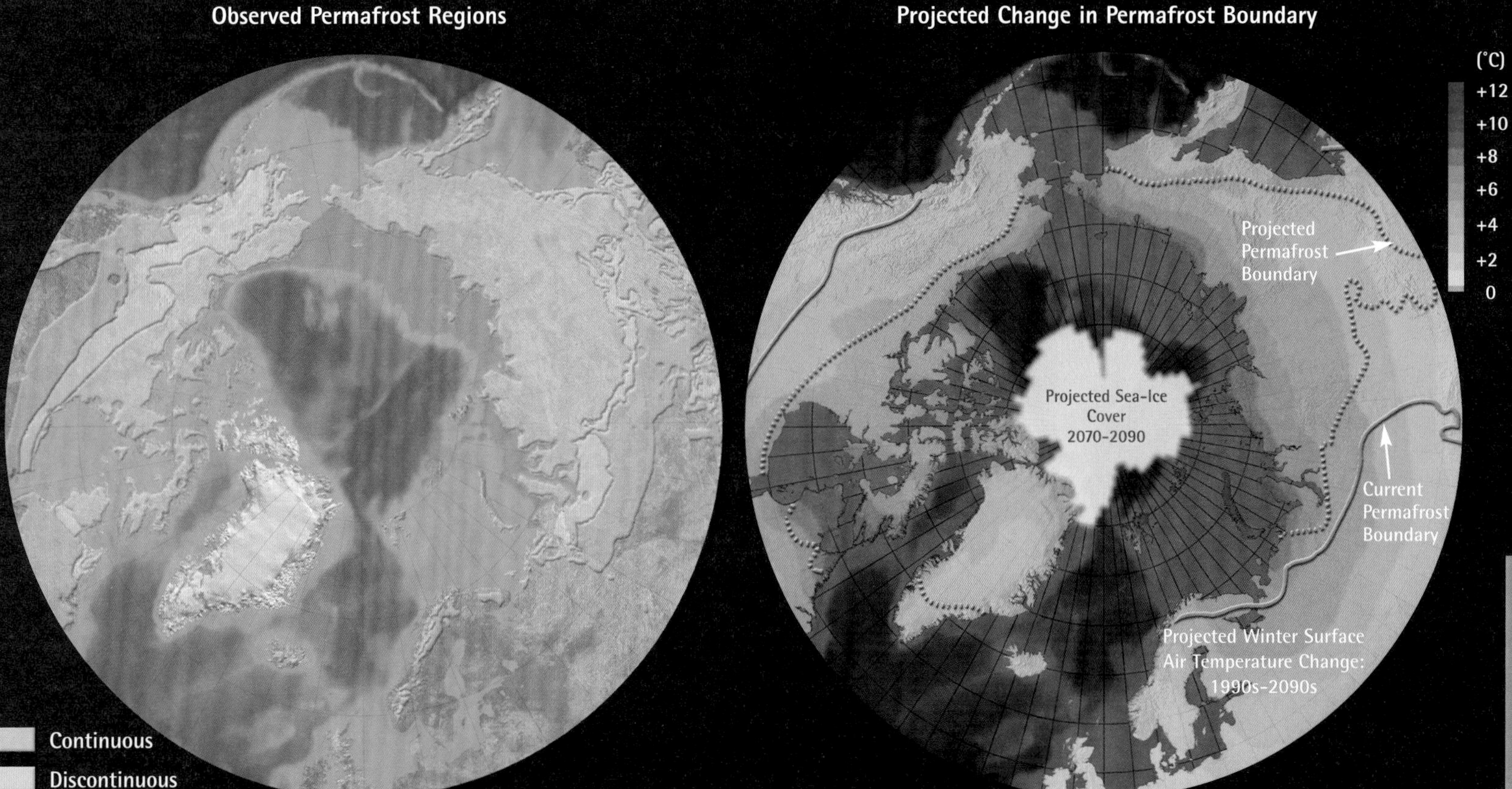

A PERMAFROST PRIMER

Permafrost is soil, rock, or sediment that has remained below 0°C for two or more consecutive years. Permafrost underlies most land surfaces in the Arctic, varying from a few meters to several hundred meters thick.

Continuous permafrost zones are those in which the permafrost occupies the entire area, and can reach up to 1500 meters in depth, for example, in parts of Siberia.

Sporadic or discontinuous permafrost zones are those in which the permafrost underlies from 10% to 90% of the land and may be only a few meters thick in places.

Active layer refers to the top layer of permafrost that thaws each year during the warm season and freezes again in winter.

Degradation of permafrost means that some portion of the former active layer fails to refreeze during winter.

Thermokarst refers to a place where the ground surface subsides and collapses due to thawing of permafrost. This can result in new wetlands, lakes, and craters on the surface.

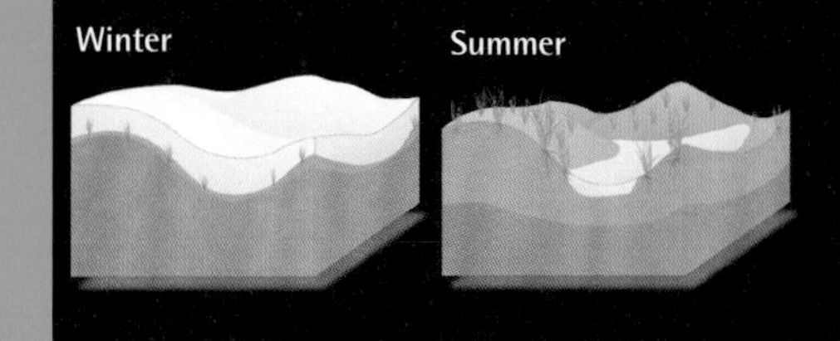

The projected rate of warming and its effects will need to be taken into account in the design of all new construction.

Impacts on Infrastructure

Projected increases in permafrost temperatures and in the depth of the active layer are very likely to cause settling, and to present significant engineering challenges to infrastructure such as roads, buildings, and industrial facilities. Remedial measures are likely to be required in many cases to avoid structural failure and its consequences. The projected rate of warming and its effects will need to be taken into account in the design of all new construction, requiring deeper pilings, thicker insulation, and other measures that will increase costs.

In some areas, interactions between climate warming and inadequate engineering are causing problems. The weight of buildings on permafrost is an important factor; while many heavy, multi-story buildings of northern Russia have suffered structural failures, the lighter-weight buildings of North America have had fewer such problems as permafrost has warmed. Continuous repair and maintenance is also required for buildings on permafrost, a lesson learned because many of the buildings that failed were not properly maintained. The problems now being experienced in Russia can be expected to occur

Infrastructure at Risk by 2050 Due to Permafrost Thaw

Stable
Low
Moderate
High

The map shows hazard potential by risk level for buildings, roads, and other infrastructure due to permafrost thaw by the middle of this century, calculated using the Hadley climate model with the moderate B2 emissions scenario. Hazard potential is partitioned into areas with high, moderate, and low susceptibility to thaw-induced settlement. Areas of stable permafrost, which are not likely to change, are also shown. A zone in the high and moderate risk category extends discontinuously around the Arctic Ocean, indicating high potential for coastal erosion. Also within these bands are population centers (Barrow, Inuvik) and river terminals on the Arctic coast of Russia (Salekhard, Igarka, Dudinka, Tiksi). Transportation and pipeline corridors traverse areas of high hazard potential in northwestern North America. The area containing the Nadym-Pur-Taz natural gas production complex and associated infrastructure in northwest Siberia also falls in the high-risk category. Large parts of central Siberia, particularly the Sakha Republic (Yakutia), and the Russian Far East show moderate or high hazard potential. Within these areas are several large population centers (Yakutsk, Noril'sk, Vorkuta), an extensive road network, and the Trans-Siberian and Baikal-Amur Mainline Railroads. The Bilibino nuclear power plant and its grid occupy an area of high hazard potential in the Russian Far East.

elsewhere in the Arctic if buildings are not designed and maintained to accommodate future warming.

Structural failures of transportation and industrial infrastructure are also becoming more common as a result of permafrost thawing in northern Russia. Many sub-grade railway lines are deformed, airport runways in several cities are in an emergency state, and oil and gas pipelines are breaking, causing accidents and spills that have removed large amounts of land from use because of soil contamination. Future concerns include a weakening of the walls of open pit mines, and pollutant effects from large mine tailing disposal facilities as frozen layers thaw, releasing excess water and contaminants into groundwater.

Building damaged due to permafrost thawing in Chersky (low Kolima River), Russia.

The effects of permafrost thawing on infrastructure in this century will be more serious and immediate in the discontinuous permafrost zone than in the continuous zone. Because complete thawing is expected to take centuries, and benefits (such as easier construction in totally thawed ground) would occur only after that time, the consequences for the next 100 years or so will be primarily negative (that is, destructive and costly).

Yakutsk, Russia Experiences Infrastructure Failure as Permafrost Thaws

In Yakutsk, a Russian city built over permafrost in central Siberia, more than 300 buildings have been damaged by thaw-induced settlement. The infrastructure affected by collapsing ground due to permafrost thaw includes several large residential buildings, a power station, and a runway at Yakutsk airport. Some ascribe a large proportion of the city's recent problems to climatic warming, though others believe that better construction practices and maintenance could have prevented much of the trouble.

BP operations center, Prudhoe Bay, Alaska, built on pilings to resist damage from thawing permafrost.

Research on the impacts of warming on infrastructure indicates that even small increases in air temperature substantially affect building stability, and that the safety of building foundations decreases sharply with increasing temperature. This effect can result in a significant decrease in the lifetime of structures as well as the potential failure of structures.

As global warming continues, detrimental impacts on infrastructure throughout the permafrost regions can be expected. Many of these impacts might be anticipated, allowing structures to be re-designed and re-engineered to withstand additional pressures under changing climatic conditions. This will certainly incur costs, but can avoid the dramatic infrastructure failures being experienced in Yakutsk and elsewhere in the Arctic.

Floods and Slides

Another set of climate-related problems for arctic infrastructure involves floods, mudslides, rockslides, and avalanches. These events are closely associated with heavy precipitation events, high river runoff, and elevated temperatures, all of which are projected to occur more frequently as climate change progresses. Soil slopes are also made less stable by thawing permafrost, and this is expected to result in more slides. Some transportation routes to markets are sensitive to the types of weather events that are expected to increase as climate continues to warm. Protecting or improving these routes will be required.

7 Thawing ground will disrupt transportation, buildings, and other infrastructure.

Impacts of Thawing Permafrost on Natural Ecosystems

Important two-way interactions exist between climate-induced changes in permafrost and vegetation. As permafrost thaws, it affects the vegetation that grows on the surface. At the same time, the vegetation, which is also experiencing impacts due to climate change, plays an important role in insulating and maintaining the permafrost. For example, forests help sustain permafrost because the tree canopy intercepts the sun's heat and the thick layer of moss on the surface insulates the ground. Thus, the projected increase in forest disturbances such as fire and insect infestations can be expected to lead to further degradation of permafrost, in addition to what is projected to result directly from rising temperatures.

Water levels in many rivers and lakes in Nunavut (eastern Arctic Canada) have been falling for four decades, with the most dramatic drops in the past decade. Native people have long relied on both drinking water and fish from these waters. Rivers also provided access by boat to hunting grounds that can no longer be reached.

In some northern forests, certain tree species (notably black spruce) utilize the ice-rich permafrost to maintain the structure of the soil in which they are rooted. Thawing of this frozen ground can lead to severe leaning of trees (sometimes referred to as "drunken forest") or complete toppling of trees. Thus, even if a longer, warmer growing season might otherwise promote growth of these trees, thawing permafrost can undermine or destroy the root zone due to uneven settling of the ground surface, leading instead to tree collapse and death. In addition, where the ground surface subsides due to permafrost thaw, even if the trees do not fall over, these sites often become the new lowest points on the landscape. At least seasonally, these places fill with water, drowning the trees.

The potential for many shallow streams, ponds, and wetlands in the Arctic to dry out under a warming climate is increased by the loss of permafrost. As permafrost thaws, ponds connect with the groundwater system. They are thus likely to drain if losses due to downward percolation and evaporation are greater than re-supply by spring snowmelt and summer precipitation. Patchy arctic wetlands are particularly sensitive to permafrost degradation that links surface waters to groundwater. Those along the southern limit of permafrost, where increases in temperature are most likely to eliminate the relatively warm permafrost, are at the highest risk of drainage. Indigenous people in Nunavut (eastern arctic Canada) have observed recently that there has been increased drying of rivers, swamps, and bogs, to the extent that access to traditional hunting grounds and, in some instances, migration of fish, have been impaired. There is also a high risk of catastrophic drainage of permafrost-based lakes, such as those found along the western arctic coast of Canada.

In other places, warming of surface permafrost above frozen ground and associated collapsing of ground surfaces could increase the formation of wetlands, ponds, and drainage networks, particularly in areas characterized by heavy concentrations of ground ice. Such thawing, however, would also lead to dramatic increases in sediment being deposited into rivers, lakes, deltas, and coastal marine environments, resulting in significant impacts to aquatic life in these bodies of water.

Changes to the water-balance of northern wetlands are especially important because most wetlands in permafrost regions are peatlands, which may absorb or emit carbon (as carbon dioxide or methane) depending on the depth of the water table. There are many uncertainties in projections of these changes. One analysis suggests that an increase in temperature of 4°C would reduce water storage in northern peatlands, even with a small and persistent increase in precipitation, causing peatlands to switch from emitting carbon dioxide to the atmosphere to absorbing it. It is also possible that the opposite could occur, whereby warming and drying could cause the rate of decomposition of organic matter to increase faster than the rate of photosynthesis, resulting in an increase in carbon dioxide emissions. A combination of temperature increase and elevated groundwater levels could result in increased methane emissions. Projections based on doubling of pre-industrial carbon dioxide levels, anticipated to occur around the middle of this century, suggest a major northward shift (by 200-300 kilometers) of the southern boundary of these peatlands in western Canada and a significant change in their structure and vegetation all the way to the coast.

"There is a lot less water, around all these islands [in Baker Lake]... There used to be a lot of water. We could go through with our outboard motors and boats, but now there is getting to be less and less water all over... There's been a lot less fish because there's not as much water anymore... The fish were more plentiful and they used to be bigger. Now you hardly get char anymore at Prince River or any of these fishing places."

L. Arngna'naaq
Baker Lake, Canada

Peatland detail

Photosynthesis
Respiration
From plants
Diffusion
From water
Respiration
Oxidation
Water table
DIC
DOC
POC
Mosses, etc.
Carbon accumulation
Permafrost
Carbon storage
Upland
Lowland
Isolated pools

Lowland

From plants
Diffusion
From water
Inflow
Oxidation
Living things
Respiration
Photosynthesis
Outflow
Diffusion
Oxidation
Sedimentation
POC
Soil
Thermocline
Respiration
Sediment and living things
Carbon accumulation
Permafrost
Peatland
DIC
DOC
POC

Carbon Cycling in Aquatic Ecosystems

Simplified schematic of the cycling of carbon in high-latitude aquatic ecosystems. Arctic wetlands typically emit carbon to the atmosphere during spring melt and as plants die in autumn. They then absorb carbon from the atmosphere as plants grow during the warm season. Future changes in the release and uptake of carbon will therefore depend on changes in vegetation, temperature, and soil conditions. Similarly, carbon cycling in lakes, ponds, and rivers will also be sensitive to direct and indirect effects of climate change.

DIC - Dissolved Inorganic Carbon
DOC - Dissolved Organic Carbon
POC - Particulate Organic Carbon

8 Indigenous communities are facing major economic and cultural impacts.

The Arctic is home to numerous Indigenous Peoples whose cultures and activities are shaped by the arctic environment. They have interacted with their environment over generations through careful observations and skillful adjustments in traditional food-harvesting activities and lifestyles. Through ways of life closely linked to their surroundings, these peoples have developed uniquely insightful ways of observing, interpreting, and responding to the impacts of environmental change.

Indigenous observations and perspectives are therefore of special value in understanding the processes and impacts of arctic climate change. There is a rich body of knowledge based on their careful observations of and interactions with their environment. Holders of this knowledge use it to make decisions and set priorities. The ACIA has attempted to combine knowledge and insights from indigenous people with data from scientific research, bringing together these complementary perspectives on arctic climate change.

Flexibility and adaptability have been key to the way arctic Indigenous Peoples have accommodated environmental changes over many generations. Current social, economic, political, and institutional changes play a part in enabling or constraining the capacity of peoples to adapt. The rapid climate change of recent decades, combined with other ongoing alterations in the world around them, presents new challenges.

Across the Arctic, indigenous people are already reporting the effects of climate change. In Canada's Nunavut Territory, Inuit hunters have noticed the thinning of sea ice, a reduction in the numbers of ringed seals in some areas, and the appearance of insects and birds not usually found in their region. Inuvialuit in the western Canadian Arctic are observing an increase in thunderstorms and lightning, previously a very rare occurrence in the region. Athabaskan people in Alaska and Canada have witnessed dramatic changes in weather, vegetation, and animal distribution patterns over the last 50 years. Saami reindeer herders in Norway observe that prevailing winds relied upon for navigation have shifted and become more variable, forcing changes in traditional travel routes. Indigenous Peoples who are accustomed to a wide range of natural climate variations are now noticing changes that are unique in the long experience of their peoples.

Compiling indigenous knowledge from across the Arctic, a number of common themes clearly emerge, though there are regional and local variations in these observations.

- The weather seems unstable and less predictable by traditional methods.
- Snow quality and characteristics are changing.
- There is more rain in winter.
- Seasonal weather patterns are changing.
- Water levels in many lakes are dropping.
- Species not seen before are now appearing in the Arctic.
- Sea ice is declining, and its quality and timing are changing.
- Storm surges are causing increased erosion in coastal areas.
- The sun feels "stronger, stinging, sharp". Sunburn and strange skin rashes, never experienced before, are becoming common.
- Climate change is occurring faster than people can adapt.
- Climate change is strongly affecting people in many communities, in some cases, threatening their cultural survival.

"The river Virma grows shallower every year. Now there is hardly any water left and it can freeze all the way to the bottom. There used to be a lot of fish, but now they are almost all gone. I think it is due to the drying of the bogs and marshes."

Vasily Lukov
Lovozero, Russia

Many indigenous communities of the Arctic depend primarily on harvesting and using living resources from the land and sea. The species most commonly harvested are marine mammals such as seals, walrus, polar bears, and narwhals, and beluga, fin, bowhead, and minke whales; land mammals such as caribou, reindeer, moose, and musk ox; fish such as salmon, Arctic char, and northern pike, and a variety of birds, including ducks, geese, and ptarmigan.

8 Indigenous communities are facing major economic and cultural impacts.

Indigenous Peoples throughout the Arctic maintain a strong connection to the environment through hunting, herding, fishing, and gathering. The living resources of the Arctic not only sustain Indigenous Peoples in an economic and nutritional sense, but also provide a fundamental basis for social identity, spiritual life, and cultural survival. Rich mythologies, vivid oral histories, festivals, and animal ceremonies illustrate the social, economic, and spiritual relationships that Indigenous Peoples have with the arctic environment. These traditions distinguish the food harvesting practices of Indigenous Peoples from conventional hunting.

Access to food resources is often related to travel access and safety. For example, changes in the rate of spring melt and increased variability associated with spring weather conditions have affected access to hunting and fishing camps. For example, when Inuit families in the western Canadian Arctic go out to camps at lakes for ice fishing and goose hunting in May, they travel by snowmobile, pulling a sled, staying on snow-covered areas or using coastal sea-ice and frozen rivers. However, warmer springs have resulted in earlier, faster snowmelt and river break-up, making access difficult. The availability of some species has changed due to the inability of people to hunt them under changing environmental conditions. For example, the reduction in summer sea ice makes ringed seals harder to find. Climate-related changes in animal distributions are occurring, and larger changes are projected. For example, northward movement of the pack-ice edge is expected to reduce the availability of seabirds as food resources to many arctic communities.

As Indigenous Peoples perceive it, the Arctic is becoming an environment at risk in the sense that sea ice is less stable, unusual weather patterns are occurring, vegetation cover is changing, and particular animals are no longer found in traditional hunting areas during specific seasons. Local landscapes, seascapes, and icescapes are becoming unfamiliar, making people feel like strangers in their own land.

Seals Become Elusive for Inuit in Nunavut, Canada

The ringed seal is the single most important food source for the Inuit, representing the majority of the food supply in all seasons. No other species is present on the land or in the waters of Nunavut in the quantities needed to sustain the dietary requirements of the Inuit. In recent decades, local people have observed that ringed seal pup production has suffered as increased temperatures have led to a reduction and destabilization of the sea ice. These ice changes have also affected the harvest of polar bear, another important food source, because ringed seals are central to a polar bear's diet and the bears are also directly affected by the observed changes in snow and ice.

To hunt, catch, and share these foods is the essence of Inuit culture. Thus, a decline in ringed seals and polar bears threatens not only the dietary requirements of the Inuit, but also their very way of life. Projections of sea-ice decline in the future spell further trouble. Forecasts of summer sea ice from climate models suggest reductions of 50% or more during this century, with some models projecting the complete disappearance of summer sea ice. Because ringed seals and polar bears are very unlikely to survive in the absence of summer sea ice, the impact on indigenous communities that depend upon these species is likely to be enormous.

The living resources of the Arctic not only sustain Indigenous Peoples in an economic and nutritional sense, but also provide a fundamental basis for social identity, spiritual life, and cultural survival.

Observed Climate Change Impacts in Sachs Harbour, Canada

The community of Sachs Harbour is located on Banks Island in the Canadian western Arctic. Climate change impacts on this community have been studied intensively through the Inuit Observations of Climate Change project, undertaken by the Community of Sachs Harbour and the International Institute for Sustainable Development. The Inuvialuit (the Inuit of the Canadian western Arctic) initiated this study because they wanted to document the severe environmental changes they are witnessing as a result of climate change and to disseminate this information to the world. A brief summary of some of their findings follows.

1. Physical Environmental Changes
- Multiyear ice no longer comes close to Sachs Harbour in summer.
- Less sea ice in summer means that water is rougher.
- Open water is now closer to land in winter.
- More rain in summer and autumn makes travel difficult.
- Permafrost is no longer solid in places.
- Lakes are draining into the sea from permafrost thawing and ground slumping.
- Loose, soft snow (as opposed to hard-packed snow) makes it harder to travel.

2. Predictability of the Environment
- It has become difficult to tell when ice is going to break-up on rivers.
- Arrival of spring has become unpredictable.
- It is difficult to predict weather and storms.
- There are "wrong" winds sometimes.
- There is more snow, blowing snow, and whiteouts.

3. Travel Safety on Sea Ice
- Too much broken ice in winter makes travel dangerous.
- Unpredictable sea-ice conditions make travel dangerous.
- Less multiyear ice means traveling on first-year ice all winter; this is less safe.
- Less ice cover in summer means rougher, more dangerous storms at sea.

4. Access to Resources
- It is more difficult to hunt seals because of a lack of multiyear ice.
- Hunters cannot go out as far in winter because of a lack of firm ice cover.
- It is harder to hunt geese because the spring melt occurs so fast.
- Warmer summers and more rain mean more vegetation and food for animals.

5. Changes in Animal Distributions and Condition
- There is less fat on the seals.
- Fish and bird species are observed that have never been seen before.
- There is an increase in biting flies; never had mosquitoes before but do now.
- Fewer polar bears are seen in the autumn because of lack of ice.
- The fish "least cisco" is now being caught in greater numbers.

8 Indigenous communities are facing major economic and cultural impacts.

Climate change is occurring faster than indigenous knowledge can adapt and is strongly affecting people in many communities. Unpredictable weather, snow, and ice conditions make travel hazardous, endangering lives. Impacts of climate change on wildlife, from caribou on land, to fish in the rivers, to seals and polar bears on the sea ice, are having enormous effects, not only for the diets of Indigenous Peoples, but also for their cultures.

The weather seems less stable and predictable.

From sources of indigenous knowledge across the Arctic come reports that the weather seems more variable, unfamiliar, and is behaving unexpectedly and outside the norm. Experienced hunters and elders who could predict the weather using traditional techniques are now frequently unable to do so. Storms often occur without warning. Wind direction changes suddenly. In many places it is increasingly cloudy. Storms bringing high winds and lightning occur with increasing frequency in some locations. As noted by several elders, "the weather today is harder to know". This presents problems for many activities, from hunting to drying fish, on which Indigenous Peoples depend.

"Right now the weather is unpredictable. In the older days, the elders used to predict the weather and they were always right, but right now, when they try to predict the weather, it's always something different."
- Z. Aqqiaruq, Igloolik, Canada, 2000

"The periods of weather are no longer the norm. We had certain stable decisive periods of the year that formed the traditional norms. These are no longer at their places... Nowadays the traditional weather forecasting cannot be done anymore as I could before... For the markers in the sky we look now in vain..."
- Heikki Hirvasvuopio, Kakslauttanen, Finland, 2002

Snow characteristics are changing, and there is more freezing rain.

Changes in snow and ice characteristics are widely reported. Changing wind patterns cause the snow to be hard packed; hunters and travel parties are thus unable to build igloos, which are still commonly relied upon for temporary and emergency shelters. Injuries and deaths have been attributed to sudden storms and those involved not being able to find good snow with which to build shelters. More freezing rain and increasing frequency of freeze-thaw cycles are affecting the ability of reindeer, caribou, musk ox, and other wildlife to find food in winter, which in turn affects the Indigenous Peoples who depend upon these animals.

"There used to be different layers of snow back then. The wind would not blow as hard, not make the snow as hard as it is now... It's really hard to make shelters with that kind of snow because it's usually way too hard right down to the ground."
- T. Qaqimat, Baker Lake, Canada, 2001

"Change has been so dramatic that during the coldest month of the year, the month of December 2001, torrential rains have fallen in the Thule region so much that there appeared a thick layer of solid ice on top of the sea ice and the surface of the land... which was very bad for the paws of our sled dogs."
- Uusaqqak Qujaukitsoq, Qaanaaq, Greenland, 2002

"It used to be that there would be proper freezing which would dry the lichen and the snow would fall on top. There would be rain that would form the bottom, which would then freeze properly. Now it rains, and the bottom freezes wet, and this is bad for the reindeers. It ruins the lichen. Ice is everywhere and the reindeer cannot get through. This has meant death to a number of reindeers because they cannot get to the lichen."
- Niila Nikodemus, 86, the oldest reindeer herder in Purnumukka, Finland, 2002

" First it snows, then it melts, like it would be summertime. And this all over again. First there is a big snowfall, then it warms up and then it freezes. During winter now it can rain, as happened last New Year. Before it never rained during wintertime. Rain in the middle of winter? To the extent that snow disappears? Yes, it is true. Rain, and snow melts!"
- Vladimir Lifov, Lovozero, Russia, 2002

The Kalaallisut (Greenlandic) word for weather and climate is sila. Sila is also the word for the universal consciousness, the all-pervading, life-giving force that is manifest in each and every person. Sila integrates and connects a person with the rhythms of the natural world.

Sea ice is declining, and its quality and timing are changing, with important repercussions for marine hunters.

Sea ice is declining markedly, both in extent and thickness. The pack ice is further from shore and often too thin to allow safe travel. Less sea ice makes stormy seas more violent and dangerous for hunters. Marine mammals whose habitat is sea ice, including walrus, polar bear, and ice-associated seals are very likely to experience major population declines in this century and could be threatened with extinction.

"Long ago, there was always ice all summer. You would see the [multiyear ice] all summer. Ice was moving back and forth this time of year. Now, no ice. Should be [multiyear]. You used to see that old ice coming from the west side of Sachs. No more. Now between Victoria Island and Banks Island, there is open water. Shouldn't be that way."
- Frank Kudlak, Sachs Harbour, Canada, 1999

"I know that today that seals, it might be because of early spring break-up or that they are out on the ice floes, that the seals are nowhere."
Man age 62, Kuujjuaq, Canada,

"When there is lots of ice, you don't worry too much about storms. You get out there and travel in between the ice [floes]. But last few years there has been no ice. So if it storms, you can't get out..."
- Andy Carpenter, Sachs Harbour, Canada, 1999

Seasonal weather patterns are changing.

Peoples across the Arctic report changes in the timing, length, and character of the seasons, including more rain in autumn and winter, and more extreme heat in summer.

"Sila [the weather and climate] has changed alright. It is a really late falltime now, and really fast and early springtime. Long ago the summer was short, but not anymore."
- Sarah Kuptana, Sachs Harbour, Canada, 1999

"It used to be really nice weather long ago when I was a kid. Bad weather now. So many mosquitoes. Sometimes it was hot, sometimes cold – not like now. [Things happen at the] wrong time now, it is way different now. August used to be cool-off time, now it is hot. It is really short in the winter now."
- Edith Haogak, Sachs Harbour, Canada, 2000

"The weather has changed to worse and to us it is a bad thing. It affects mobility at work. In the olden days the permanent ice cover came in October... These days you can venture to the ice only beginning in December. This is how things have changed."
- Arkady Khodzinsky, Lovozero, Russia, 2002

I think over again
My small adventures
My fears
Those small ones that seemed so big
For all the vital things
I had to get and to reach
And yet there is only one great thing
To live to see the great day that dawns
And the light that fills the world.

- Inuit Poem

9 Elevated ultraviolet radiation levels will affect people, plants, and animals.

The ozone layer absorbs UV from the sun, protecting life on earth from exposure to excessive levels of UV. Ozone depletion may thus lead to increases in UV levels at the earth's surface.

Ozone in the Atmosphere

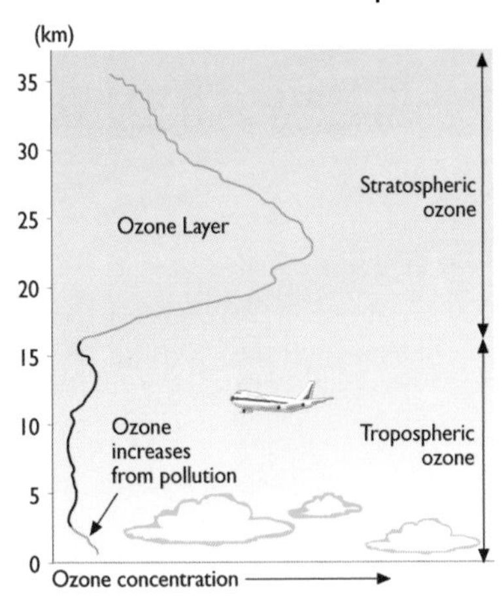

Most ozone resides in the stratosphere, relatively high above the earth's surface, where it protects life on earth from excess UV radiation. Increases in ozone levels occur near earth's surface as a result of pollution. This ground-level ozone, also known as smog, causes respiratory problems in humans and other negative impacts. The discussion in this report concerns stratospheric ozone, not ground-level ozone.

Ultraviolet radiation (UV) reaching the earth's surface is a growing concern in the Arctic, largely due to depletion of stratospheric ozone caused by emissions of chlorofluorocarbons (CFCs) and other manmade chemicals over the last 50 years. Ozone depletion over the Arctic has been severe and is greatest in the spring when living things are most vulnerable.

While the international treaty known as the Montreal Protocol (and subsequent amendments that strengthened it) has phased out the production of most of these ozone-destroying chemicals, many have long atmospheric lifetimes and so those previously released will continue to destroy ozone for decades to come. Ozone depletion in the Arctic is highly sensitive to changes in temperature, meaning that ozone levels are likely to be strongly influenced by climate change, even though the fundamental depletion processes involve ozone-destroying chemicals produced by human activities.

Although the uncertainty in future ozone projections is high, the long timeframe for ozone recovery suggests that the Arctic is very likely to be subject to elevated levels of UV for several decades. Increased UV levels are likely to affect many living things in the Arctic. In humans, excess levels are known to cause skin cancer, sunburn, cataracts, cornea damage, and immune system suppression. Ultraviolet radiation is also known to cause or accelerate damage to a number of materials used in the region's infrastructure. There are also likely to be wide-ranging impacts on natural ecosystems.

Many people confuse the issues of ozone depletion and climate change. While the two are related in a number of ways, they are driven by two distinct mechanisms. Human-induced climate change results from the build up of carbon dioxide, methane, and other greenhouse gases that trap heat in the lower atmosphere (called the troposphere), causing global warming. Ozone depletion results from the human-induced build-up of chlorinated chemicals such as byproducts of CFCs and halons that break apart ozone molecules through chemical reactions that take place in the stratosphere.

The United Nations Environment Programme (UNEP) and the World Meteorological Organization (WMO) have undertaken periodic assessments of changes in stratospheric ozone and ultraviolet radiation. The most recent UNEP/WMO Scientific Assessment of Ozone Depletion was completed in 2002. The ACIA has built upon and extended the findings of that assessment.

Arctic Ozone Depletion

The ozone layer absorbs UV from the sun, protecting life on earth from exposure to excessive levels of UV. Ozone depletion may thus lead to increases in UV levels at the earth's surface. The most severe depletion has taken place in the polar regions, causing the so-called Antarctic "ozone hole," and a similar, though less severe, seasonal depletion over the Arctic. Varying degrees of depletion extend around the globe, generally becoming less severe with increasing distance from the poles.

The accumulated annually averaged loss of ozone over the Arctic has been about 7% since 1979. But this obscures much larger losses at particular times of the year and on particular days, and it is these losses that have the potential for significant biological impacts. The largest ozone reductions have occurred in spring, with average springtime losses of 10-15% since 1979. The largest monthly deviations, 30-35% below normal, were in March 1996 and 1997. Daily ozone values were 40-45% below normal in March-April 1997. Major ozone losses (defined as greater than 25% depletion) lasting several weeks have been observed during seven of the past nine springs in the Arctic.

Factors Determining UV at the Surface

Ozone levels directly influence the amount of UV reaching the earth's surface. Surface UV levels are also strongly affected by clouds, the angle of the sun's rays, altitude, the presence of tiny particles in the atmosphere (which scientists refer to as aerosols), and the reflectivity of the surface (determined largely by the extent of snow cover, which is highly reflective). These factors change from day to day, season to season, and year to year, and can increase or decrease the amount of UV that reaches living things at the surface. The highest doses of UV in the Arctic are observed in the spring and summer, due primarily to the relatively high angle of the sun. The low sun angle during autumn and winter creates a great deal of diffuse UV scattered from the atmosphere and reflected off snow and ice. Reflectance off snow can increase the dose received by living things at the surface by over 50%.

UV Protection by Ozone Layer

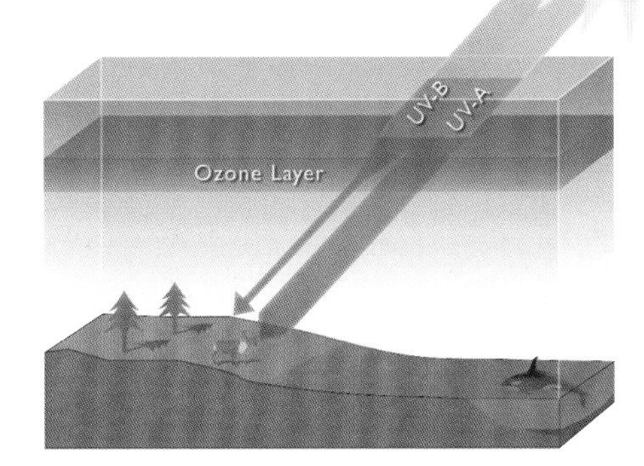

The stratospheric ozone layer absorbs some of the ultraviolet radiation from the sun. UV-B radiation is most strongly absorbed by ozone, greatly reducing the amount that reaches the earth. UV-A and other types of solar radiation are not strongly absorbed by ozone. Human exposure to UV increases the risk of skin cancer, cataracts, and a suppressed immune system. UV exposure can also damage plant and animal life on land, in the oceans, and in rivers and lakes.

Key Finding #9

The various factors affecting UV doses can have multi-faceted effects, some of which are likely to be influenced by climate change. For example, snow and ice reflect solar radiation upward, so plants and animals on top of the ice are likely to receive lower doses as snow and ice recede due to warming. On the other hand, plants and animals below the snow and ice, which were previously protected by that cover, will receive more UV as snow and ice recede. The projected reduction in snow and ice cover on the surface of rivers, lakes, and oceans is thus likely to increase the exposure of many living things in these water bodies to damaging levels of UV. In addition, the projected earlier spring melting of snow and ice cover comes at the time of year when UV radiation is most likely to be elevated due to ozone loss.

As Ozone Declines, UV Rises

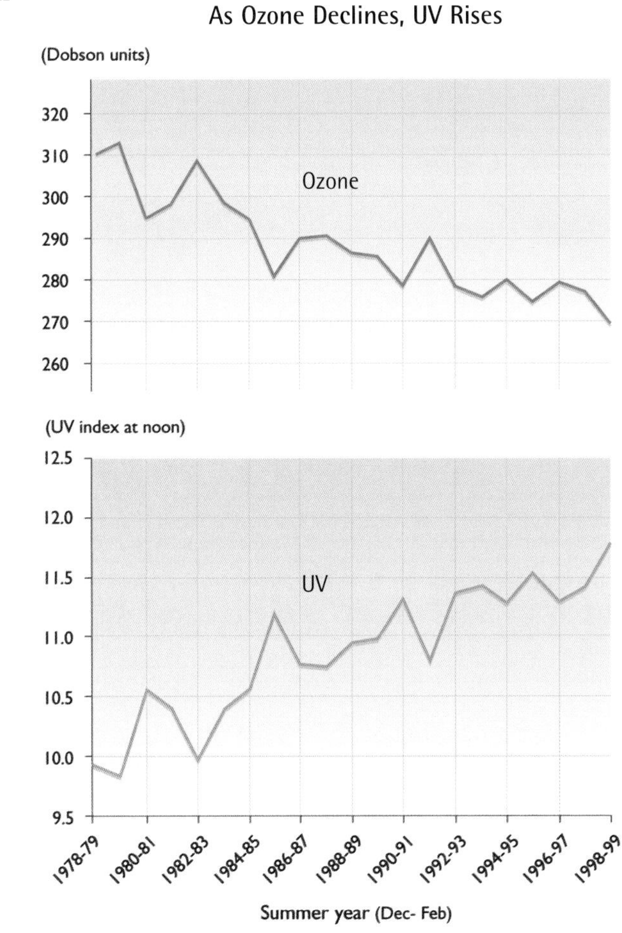

This graph demonstrates the well-established fact that all other factors being equal, less stratospheric ozone results in more UV radiation at the surface. Other factors also affect UV levels, including clouds, snow, and ice, and any of these can alter this simple relationship.

Factors Affecting UV at the Surface

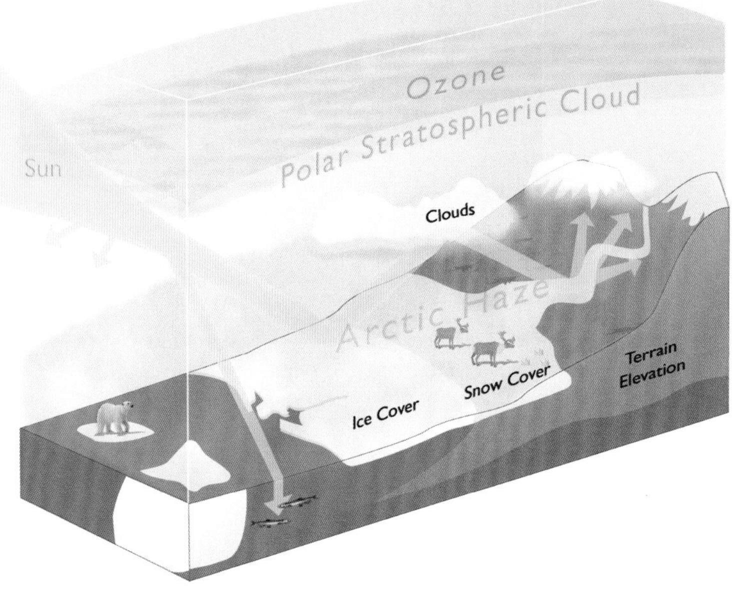

Ozone levels, clouds, the angle of the sun's rays, altitude, tiny particles in the atmosphere (which scientists refer to as aerosols), and the reflectivity of the surface (determined largely by the extent of snow cover, which is highly reflective), all influence the amount of UV reaching the surface.

9 Elevated ultraviolet radiation levels will affect people, plants, and animals.

Variations Over Time and Space

The ozone depletion and resulting increase in UV levels at the surface in the Arctic have not been symmetrical around the pole. Depletion is also not consistent over time; some years show strong depletion while others do not, due to variations in the dynamics and temperature of the stratosphere. There is a great deal of natural variability in ozone levels, and along with the long-term changes due to human activities, natural variations continue to occur. Transport of ozone can result in days of very high or very low UV levels. Because of the nature of ozone depletion, elevated UV is generally observed in spring, when biological systems are most sensitive. Increased UV doses, especially when combined with other environmental stresses, pose a threat to some arctic species and ecosystems.

Because ozone depletion is expected to persist over the Arctic for several more decades, model results anticipate up to a 90% increase in spring UV doses for 2010-2020, relative to those in 1972-1992.

Arctic Ozone Recovery Delayed by Climate Change

No significant improvement in stratospheric ozone levels over the Arctic is projected for the next few decades. One reason is that increasing levels of greenhouse gases, while warming the troposphere, actually cool the stratosphere. This can worsen ozone depletion over the poles because lower temperatures strengthen the swirl of winds known as the polar vortex and encourage the formation of polar stratospheric clouds. The icy particles of these clouds are sites on which ozone-destroying chemical reactions occur. And the vortex isolates the stratosphere over the Arctic and prevents ozone from outside the region from replenishing the depleted ozone over the Arctic. Thus, for the next few decades, ozone depletion and elevated UV levels are projected to persist over the Arctic. At the same time, a reduction in springtime snow and ice cover due to warming is likely to expose vulnerable young plant and animal life to elevated UV levels.

Arctic Ozone in March

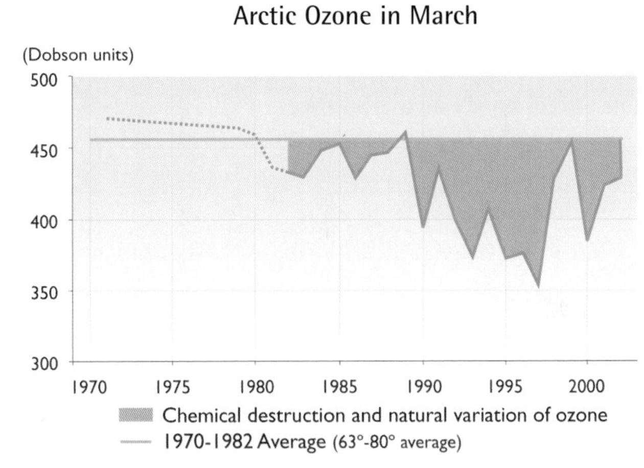

Ozone levels vary significantly from year to year. There is also a strong downward trend in ozone that is especially pronounced at the poles. This graph shows a pre-depletion average (solid red line) compared to ozone levels in more recent years. Natural variations in meteorological conditions influence the year-to-year changes, especially in the Arctic where depletion is highly sensitive to temperature. The blue line represents a monthly average in March in the Arctic. After 1982, significant depletion is found in most years.

Because ozone depletion is expected to persist over the Arctic for several more decades, episodes of very low spring ozone levels are likely to continue. Model results anticipate up to a 90% increase in springtime UV doses for 2010-2020 relative to those in 1972-1992. Because the models used to make these projections assume full compliance with the Montreal Protocol and its amendments, ozone recovery is likely to be slower and UV levels higher than projected if the phase-out of ozone-depleting chemicals is not achieved as outlined by the Protocol and its amendments.

Polar stratospheric clouds.

No significant improvement in stratospheric ozone levels over the Arctic is projected for the next few decades.

Key Finding #9

Layers of the Atmosphere

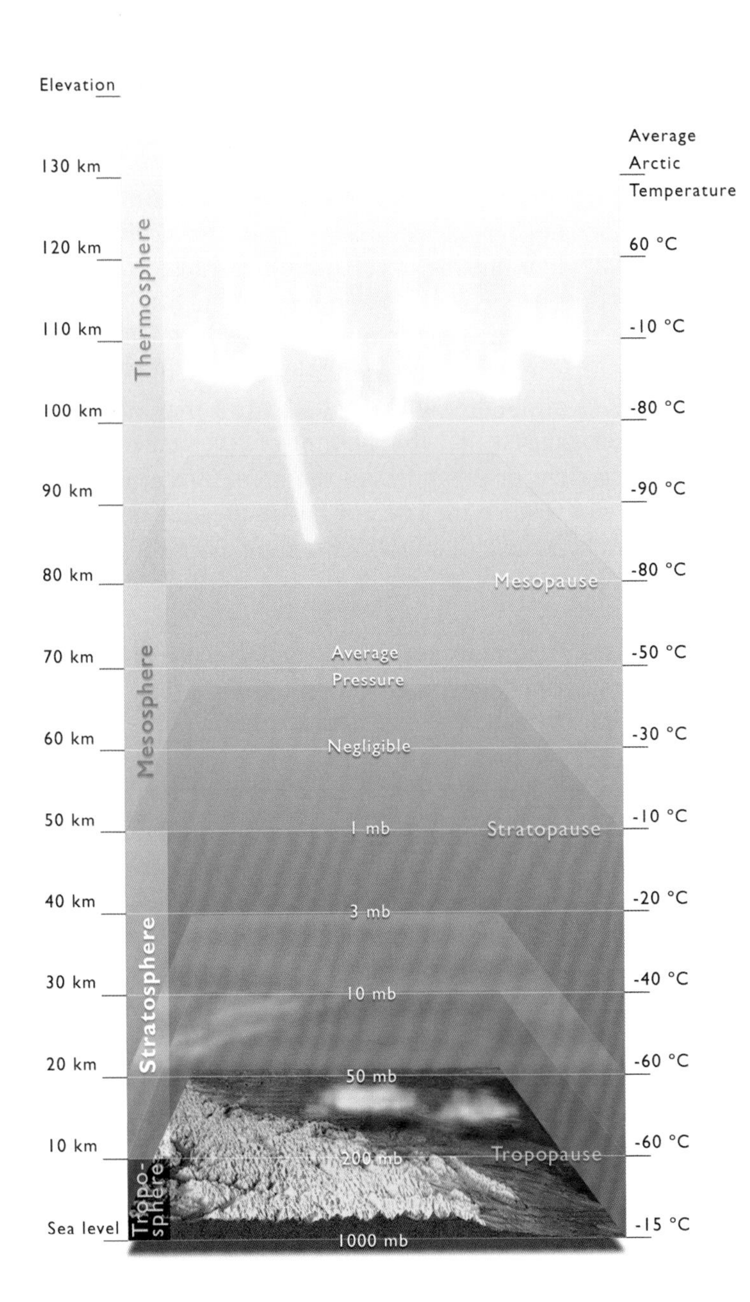

Temperatures in the Polar Lower Stratosphere

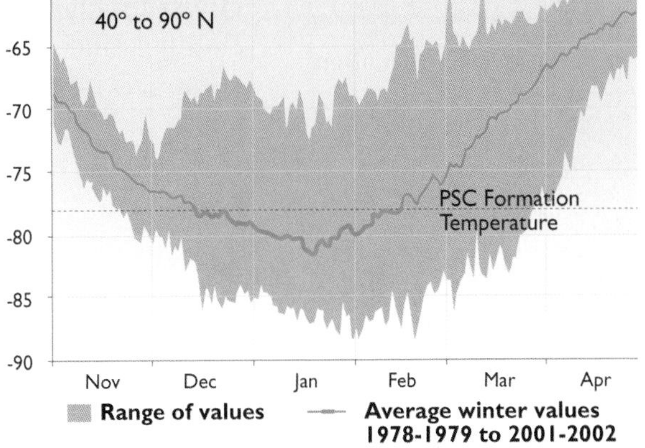

Over the Arctic, minimum air temperatures in the lower stratosphere are near -80˚C in January and February. Polar stratospheric clouds (PSCs) form when temperatures fall below -78˚C. The icy particles of these clouds are sites upon which ozone-destroying chemical reactions take place. Increasing concentrations of greenhouse gases, while warming the air near the earth's surface, act to cool the stratosphere, causing these clouds to form for a longer period of time, worsening ozone depletion.

9 Elevated ultraviolet radiation levels will affect people, plants, and animals.

Impacts of UV on People

Human beings receive about half their lifetime UV dose by the time they are 18 years old. Current elevated UV levels in the Arctic indicate that the present generation of young people is likely to receive about a 30% greater lifetime UV dose than any prior generation. Such increases in UV doses are important to people of the Arctic because UV can induce or accelerate incidence of skin cancer, cornea damage, cataracts, immune system suppression, viral infections, aging of the skin, sunburn, and other skin disorders. Skin pigmentation, while protecting to some extent against skin cancer, is not an efficient protector against UV-induced immune system suppression. The immunosuppressive effects of UV play an important role in UV-induced skin cancer by preventing the destruction of skin cancers by the immune system. Some evidence suggests a connection between exposure to sunlight and non-Hodgkin's lymphoma and autoimmune diseases such as multiple sclerosis, with the relationship suggested to be via the immunosuppressive effects of UV. Because UV is known to activate viruses such as the herpes simplex virus through immune suppression, increased UV could increase the incidence of viral diseases among arctic populations, particularly as climate warming may introduce virus-carrying insect species to the Arctic.

Eye damage is of particular concern in the Arctic. UV has traditionally been measured on a flat, horizontal surface, but this does not represent the way a person receives a UV dose. People, who are generally vertical when outside, receive a higher dose than a horizontal surface, largely due to reflection from snow. Measurements incorporating this fact indicate that springtime ozone depletion could contribute greatly to UV effects on the eyes due to the significance of snow reflection. Observations show that UV doses to vertical surfaces such as the eyes are higher at the end of April than at any other time of the year. These high doses suggest that the amount of UV received when looking toward the horizon can be equivalent or greater than the amount received when looking directly upward. People can reduce the risks of UV-induced health effects by limiting their exposure through the use of sunscreens, sunglasses, protective clothing, and other preventative measures.

Changes in Surface UV Radiation

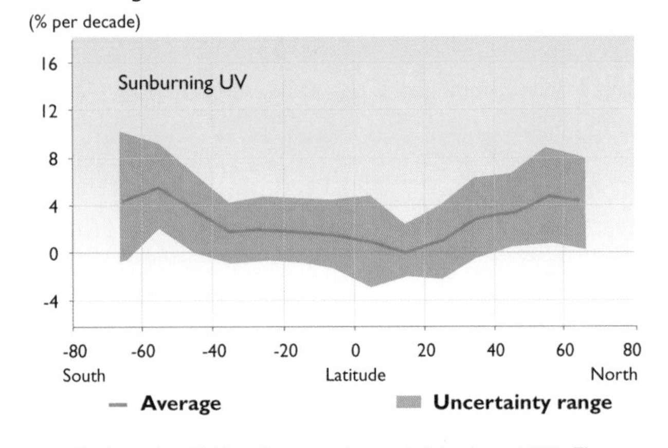

Sunburning UV has increased worldwide since 1980. The graph shows this UV at the surface, estimated from observed decreases in ozone and the relationship between ozone and UV established at various locations. The UV increases are largest near the poles because the reduction in ozone has been largest there.

In addition to the impacts on human health, UV radiation is known to adversely affect many materials used in construction and other outdoor applications. Exposure to UV can alter plastics, synthetic polymers used in paint, and natural polymers present in wood. Increased UV exposure due to ozone depletion is therefore likely to decrease the useful life of these materials and add costs for more frequent painting and other maintenance. High surface reflectivity due to snow cover and long hours of sunlight in spring and summer along with springtime ozone losses can combine to deliver a high cumulative UV dose to vertical surfaces such as the walls of buildings, leading to degradation of susceptible materials. The high winds and repeated freezing and thawing that occur in the Arctic may exacerbate materials problems that can develop as a result of UV damage. The costs of early replacement imply rising infrastructure costs that are likely to be paid for by individuals.

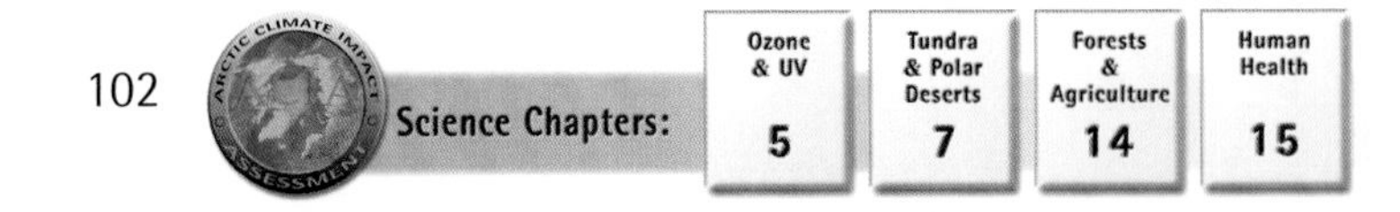

Impacts of UV on Ecosystems

Ecosystems on Land

Plants and animals show a variety of effects from increased UV radiation, though these effects vary widely by species. In the short term, a few species are projected to benefit, while many more would be adversely affected. Long-term effects are largely unknown. In addition to direct effects, animals will be indirectly affected by changes in plants. For example, pigments that are needed to protect plants against UV also make them less digestible for the animals that depend upon them. So while some plants can adapt to higher UV levels by increasing their pigmentation, there are often wider implications of this adaptation for dependent animals and ecosystem processes. Increased UV also has long-term impacts on ecosystem processes that reduce nutrient cycling and can decrease productivity.

Springtime in the Arctic is a critical time for the birth and growth of animals and plants. Historically, ozone had been at its highest levels in the spring, offering living things the heightened UV protection they needed during this sensitive time. Since ozone depletion due to manmade chemicals became a problem several decades ago, spring is now the time of year with the largest losses of stratospheric ozone. Longer daylight hours in springtime also add to UV exposure. Increases in UV also interact with climate change, such as the warming-related decrease in springtime snow cover, creating the potential for increased impacts on plants, animals, and ecosystems.

Birch Forests at Risk from Impacts of UV and Warming on the Autumn Moth

One example of a documented impact of increased UV that also has interactions with climate warming involves the autumn moth, an insect that eats the leaves of birch trees, causing tremendous damage to forests. Increased UV modifies the chemical structure of the birch leaves, greatly reducing their nutritional value. The moth caterpillars thus eat up to three times more than normal to compensate. Increased UV also appears to improve the immune system of the autumn moth. In addition, UV destroys the polyhydrosis virus that is an important controller of the survival of moth caterpillars. Increased UV is thus expected to lead to increased caterpillar populations that would in turn lead to more birch forest defoliation. At the same time, winter temperatures below -36˚C have previously limited the survival of autumn moth eggs, controlling moth populations. When winter temperatures rise above that threshold, caterpillar survival increases. Thus, observed and projected winter warming is expected to further increase moth populations, thus increasing damage to birch forests. The damaging impacts of climate change are likely to exceed the impacts of UV on birch forests.

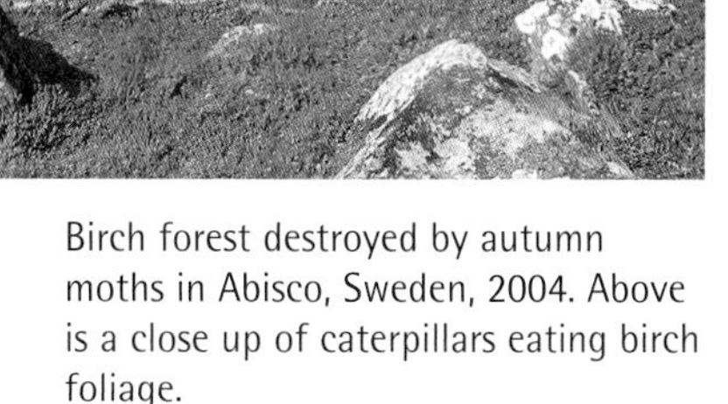

Birch forest destroyed by autumn moths in Abisco, Sweden, 2004. Above is a close up of caterpillars eating birch foliage.

9 Elevated ultraviolet radiation levels will affect people, plants, and animals.

Freshwater Ecosystems

Some freshwater species, such as amphibians, are known to be highly sensitive to UV radiation, though the vulnerability of northern species has been little examined. Climate-related changes are projected in three important factors that control the levels of UV that reach living things in freshwater systems: stratospheric ozone, snow and ice cover, and materials dissolved in water that act as natural sunscreens against UV. Reduced stratospheric ozone is expected to persist for several decades, allowing increased UV levels to reach the surface, particularly in spring.

More significantly for aquatic life, the warming-induced reduction in springtime snow and ice cover will decrease protection for plants and animals normally shielded by that cover, leading to major increases in underwater UV exposure. White ice and snow form significant barriers to UV penetration; just two centimeters of snow can reduce the below-ice exposure to UV by about a factor of three. This is especially important in freshwater systems that contain low levels of dissolved matter that would shield against UV.

Climate warming is expected to increase levels of dissolved matter in many arctic freshwater systems as warming increases vegetation growth.

Lakes and ponds in northern areas of the Arctic generally contain much less dissolved material than those in the southern part of the region, due mainly to the greater vegetation that surrounds water bodies in the south. Arctic waters also contain little aquatic vegetation. In addition to the low levels of dissolved matter and resulting deep penetration of UV in arctic lakes and ponds, many of these freshwater systems are quite shallow. For example, the average depth of more than 900 lakes in northern Finland and about 80 lakes in arctic Canada is less than 5 meters. As a consequence, all living things, even those at the bottom of the lakes, are exposed to UV radiation.

Effects of Increased UV In Freshwater Ecosystems

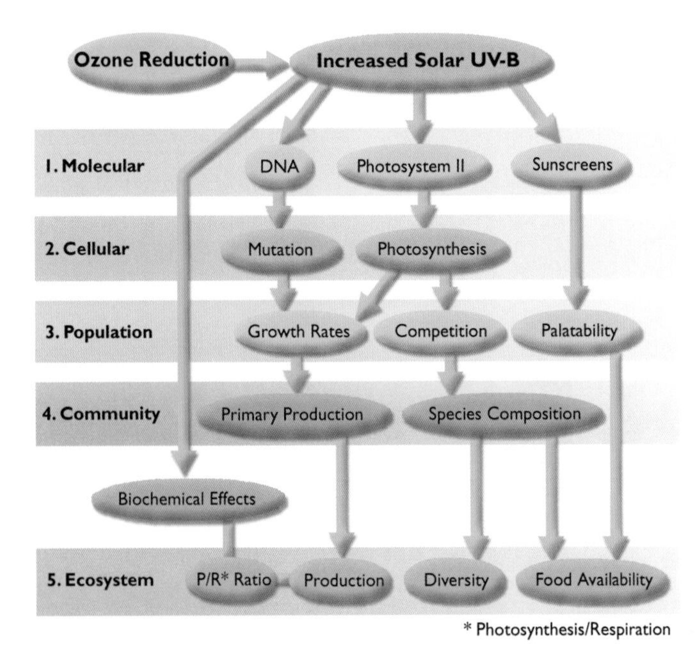

UV is the most reactive waveband in the solar spectrum and has a wide range of effects, from the molecular level to the level of the whole ecosystem.

Some of the first impacts of warming will be associated with the loss of permanent ice cover in far northern lakes; these impacts are already taking place in the Canadian High Arctic. As the length of the ice-free season increases, these effects will be amplified. However, climate warming is expected to increase levels of dissolved matter in many arctic freshwater systems as warming increases vegetation growth. In addition, thawing permafrost could increase the amount of sediment stirred up in the water, adding protection against UV. These changes could partially offset the increases in UV due to reduced snow and ice cover and to decreased ozone levels.

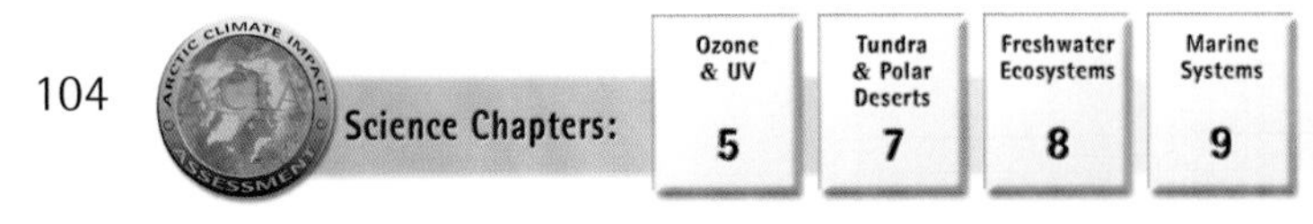

Marine Ecosystems

Phytoplankton, the tiny plants that are the primary producers of marine food chains, can be negatively impacted by exposure to UV radiation. Severe UV exposure can decrease productivity at the base of the food chain, perhaps by 20-30%. Current levels of UV negatively affect some secondary producers of marine food chains; UV-induced deaths in early life stages and reduced survival and ability to reproduce have been observed. Damage to the DNA of some species in samples collected from depths of up to 20 meters has been detected. Some species suffer strong negative impacts while others are resistant, depending on season and location of spawning, presence of UV screening substances, ability to repair UV-induced damage, and other factors.

There is clear evidence of detrimental effects of UV on early life stages of some marine fish species. For example, in one experiment, exposure to surface levels of UV killed many northern anchovy and Pacific mackerel embryos and larvae; significant sub-lethal effects were also reported. Under extreme conditions, this experiment suggested that 13% of the annual production of northern anchovy larvae could be lost. Atlantic cod eggs in shallow water (50 centimeters deep) also show negative effects due to UV exposure.

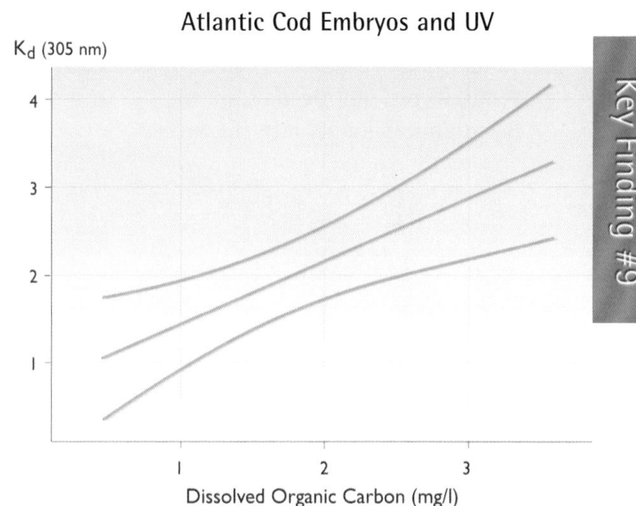

Atlantic cod embryos are sensitive to UV radiation. If protected from UV exposure, whether by stratospheric ozone, clouds, or dissolved organic carbon, their survival improves sharply. This graphic illustrates the level of protection provided by the organic matter content of the water column, with survival improving with increasing levels of dissolved organic matter. Climate change could affect levels of dissolved matter in water.

UV-induced changes in food chain interactions are likely to be more significant than direct effects on any one species. For example, UV exposure, even at low doses, reduces the content of important fatty acids in algae, decreasing the levels of these essential nutrients available to be taken up by fish larvae. Since fish larvae and the chain of predators through the food web require these essential fatty acids for proper development and growth, such a reduction in the nutritional quality of the food base has potentially widespread and significant implications for the overall health and productivity of the marine ecosystem. Exposure to UV radiation has many harmful effects on the health of fish and other marine animals, notably the suppression of the immune system. Even a single UV exposure decreases a fish's immune response, and the reduction is still visible 14 days after the exposure. This could cause increased susceptibility to disease by whole populations. The immune systems of young fish are likely to be even more vulnerable to UV as they are in critical stages of development, resulting in compromised immune defenses later in life.

Recent studies estimate that a 50% seasonal reduction in stratospheric ozone could reduce primary production in marine systems by up to 8.5%. However, as with freshwater systems, cloud cover, ice cover, and the clarity or opaqueness of the water will also be important factors in determining UV exposure.

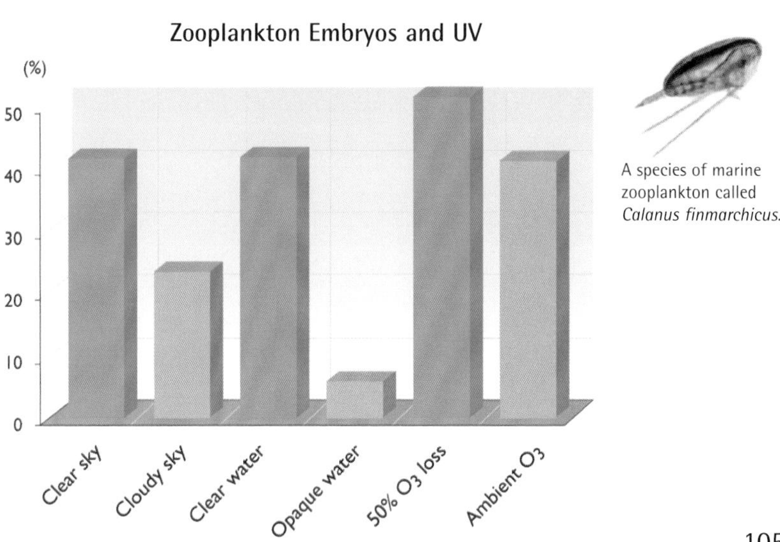

Model simulations of the relative effects of selected variables on UV-induced death in *Calanus finmarchicus*. The graph illustrates that clouds, water opacity, and ozone all reduce embryo deaths due to UV, but that the opacity of the water column has the strongest protective effect of the three variables. Zooplankton are an essential part of the marine food chain.

A species of marine zooplankton called *Calanus finmarchicus*.

Key Finding #9

10 Multiple influences interact to cause impacts to people and ecosystems.

Climate change in the Arctic is taking place within the context of many other changes, such as chemical pollution, increased ultraviolet radiation, and habitat destruction. Societal changes include a growing population, increasing access to arctic lands, technological innovations, trade liberalization, urbanization, self-determination movements, increasing tourism, and more. All of these changes are interrelated and the consequences of these phenomena will depend largely on interactions among them. Some of these changes will exacerbate impacts due to climate change while others alleviate impacts. Some changes will improve peoples' ability to adapt to climate change while others hinder their adaptive capacity.

The degree to which people are resilient or vulnerable to climate change depends on the cumulative stresses to which they are subject as well as their capacity to adapt to these changes. Adaptive capacity is greatly affected by political, legal, economic, social, and other factors. Responses to environmental changes are multi-dimensional. They include adjustments in hunting, herding, and fishing practices as well as alterations in the political, cultural, and spiritual aspects of life. Adaptation can involve changes in knowledge and how it is used, for example, using newfound knowledge of weather and climate patterns. People can alter their hunting and herding grounds and the species they pursue, and build new partnerships between federal governments and Indigenous Peoples' governments and organizations.

The particular environmental changes that create the greatest stresses vary among arctic communities. For example, threats to human health from persistent organic pollutants (POPs) and the reduction in sea ice are extremely serious for Inuit in northern Canada and western Greenland, but not as important to Saami in northern Norway, Sweden, and Finland. For the Saami, freezing rain that coats reindeer forage with ice is of great concern, as is the encroachment of roads on grazing lands.

Climate Change and Contaminants

Contaminants including POPs and heavy metals transported to the Arctic

Wind, Rivers, and Ocean Currents Bring Contaminants into the Arctic

Gyre
Warm Currents
Cold Currents
River Outflows
Catchment Area for Arctic
Wind Flow

Contaminants emitted in northern industrial areas are transported to the Arctic where they may become concentrated as they move up the food chain.

from other regions are among the major environmental stresses that interact with climate change. Certain arctic animal species, particularly those high on the marine food chain, carry high levels of POPs such as DDT and PCBs. Global use of these chemicals peaked in the 1960s and 1970s and their manufacture has since been banned in most countries. However, pollutants emitted prior to these controls persist in the environment and are transported, primarily by air currents, from industrial and agricultural sources in the mid-latitudes to the Arctic where they condense out of the air onto particles or snowflakes or directly onto earth's surface.

POPs become increasingly concentrated as they move up the food chain, resulting in high levels in polar bear, arctic fox, and various seals, whales, fish, seabirds, and birds of prey. Arctic people who eat these species are thus exposed to potentially harmful levels of these pollutants. Levels of concern have been measured in blood samples from people in various arctic communities, for example, in eastern Canada, Greenland, and eastern Siberia, with strong variations observed around the region.

Mercury is the heavy metal of greatest concern in parts of the Arctic. Mercury from distant sources is deposited onto snow in the Arctic where it is released to the environment when the snow melts in springtime, at the onset of animal and plant reproduction and rapid growth, when living things are most vulnerable. Coal burning, waste incineration, and industrial processes are the major sources of global mercury emissions. Current mercury levels pose a health risk to some arctic people and animals, and because mercury is so persistent, mercury levels are still increasing in the region, despite emissions reductions in Europe and North America.

Winds carry contaminants, and precipitation deposits them onto the land and sea. Temperature plays a role in determining the distribution of contaminants between air, land, and water. Projected climate change-related alterations in wind patterns, precipitation, and temperature can thus change the routes of contaminant entry and the locations and amounts of deposition in the Arctic. More extensive melting of multi-year sea ice and glaciers results in the rapid release of large pulses of pollutants that were captured in the ice over years or decades.

There are several other ways that climate change can alter contaminant pathways into the Arctic. Recent evidence suggests that salmon migrations undergo large, climate-related variations and that Pacific salmon may respond to change by moving northward into arctic rivers. These salmon accumulate and magnify contaminants in the Pacific Ocean, and transport them into arctic waters. For some lakes, fish may bring in more POPs than does atmospheric deposition. Similarly, changing bird migrations have the potential to transport and concentrate contaminants in particular watersheds. For example, Norwegian researchers studying Lake Ellasjoen found that seabirds serve as an important pathway for contaminants (in this case POPs) from marine to freshwater environments.

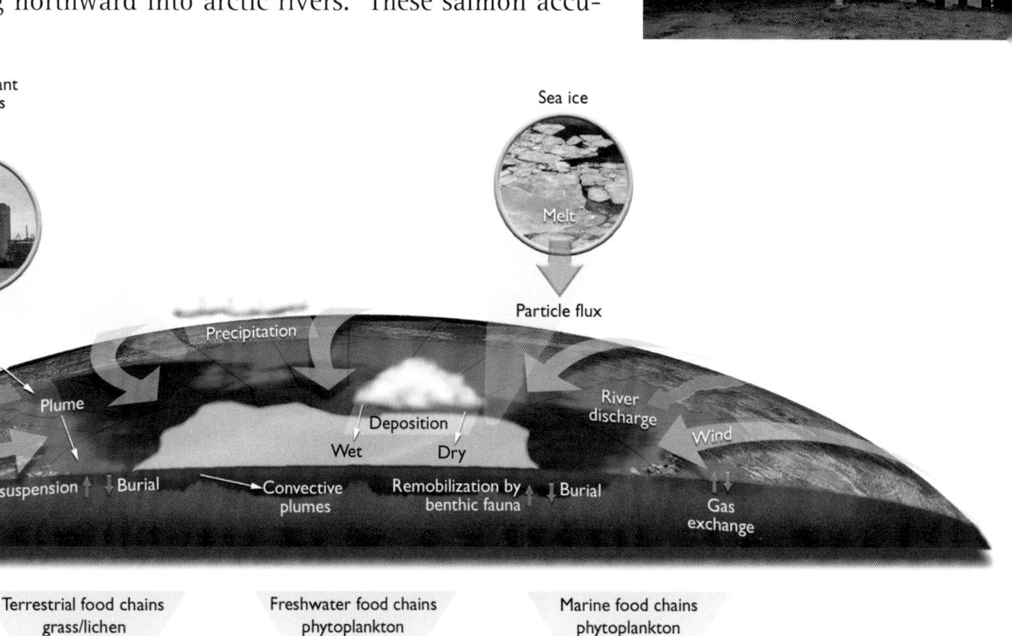

10 Multiple influences interact to cause impacts to people and ecosystems.

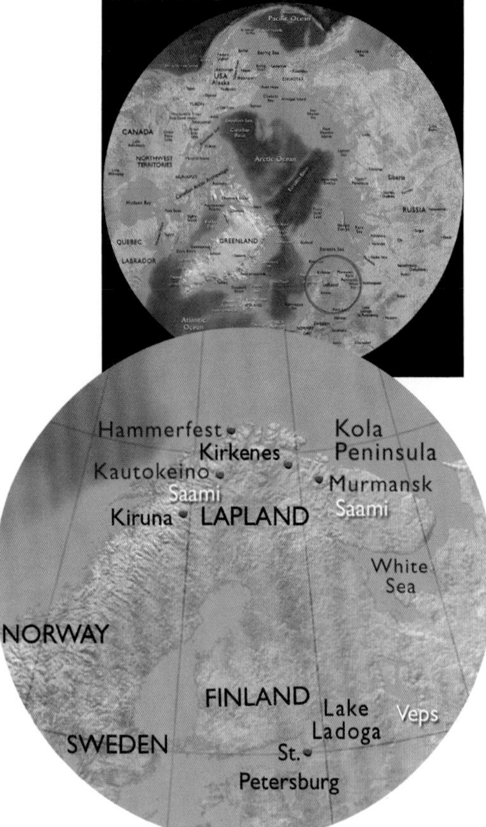

Case Study of Interacting Changes: Saami Reindeer Herders

Observed and projected increases in temperature and precipitation and changes in the timing of the seasons affect reindeer herding in numerous ways. Increases in the frequency of rain on snow, and in periods of winter melting, result in the formation of ice crust layers that make forage less accessible. Increasing autumn temperatures might lead to a later start of the period with snow cover. Rising temperatures and precipitation could increase the frequency of snow falling on unfrozen ground. An increased number, density, and distribution of birch trees in grazing areas has already begun to decrease the availability of forage plants for reindeer in winter. Shifts of forest vegetation into tundra areas are likely to further reduce traditional pasture areas.

The characteristic seasonal pattern of moving herds between winter and summer pastures reflects the herders' knowledge of seasonal changes in the availability of key resources such as forage and water. In the warm winters of the 1930s, for example, when conditions were sometimes difficult owing to heavy precipitation, herds were moved to the coast earlier than normal in the spring. Similarly, the movement of herds from poorer to better grazing areas, including the "trading of good snow" by neighboring herders, reflects thorough knowledge of forage conditions. In every case, the success of the herders is contingent upon the freedom to move.

A variety of factors, including government policies in the past few decades, have constrained the ability of Saami reindeer herders to respond to and cope with climate warming and other changes. One important stress has come from the encroachment of roads and other infrastructure on traditional reindeer grazing lands. Another stress comes from conflicting objectives among parties. Norway's mountain pastures are an important resource for herders, but pastureland management is complicated by the

"The world has changed too much now. We can say nature is mixed up now. An additional factor is that reindeer herding is being pressured from different political, social, and economic fronts at all times now. Difficulties are real. A way of living that used to support everything is now changing."

Veikko Magga
Saami Reindeer Herder
Vuotso, Finland

presence of predators such as lynx, wolf, and wolverine, which are a major threat to the survival of reindeer calves, but are protected by wildlife conservation efforts.

Other changes come from laws that emphasize meat production, encouraging active breeders and discouraging small herds. These laws favor larger herds, which have thus increased from around 100 to 700 animals. These laws also favor herds dominated by females and calves (the calves are slaughtered for meat) and have resulted in a change in structure from a traditional herd consisting of about 40% bulls, to herds with only 5% bulls. In traditional Saami herding practices, the bulls are important because their superior ability to dig through deep or poor quality snow make forage plants available to the entire herd. The reduced proportion of bulls may become more of a problem in the future if snow conditions altered by climate change make grazing even more difficult for smaller reindeer.

Road Expansion Reduces Reindeer Pasture

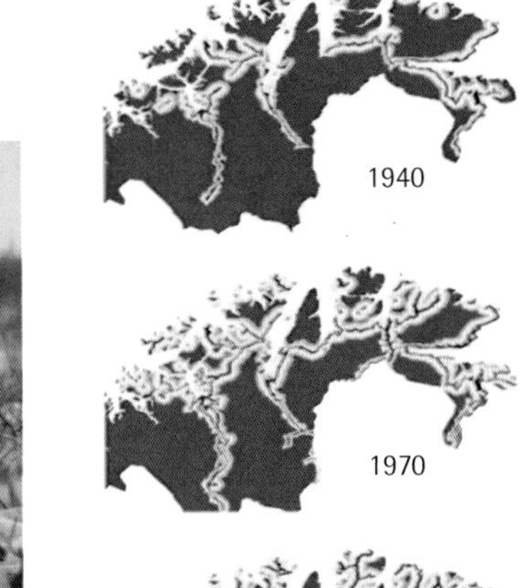

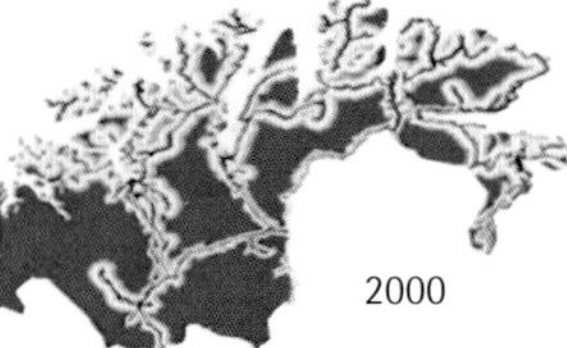

Impact
(Reduced abundance of wildlife)

- Very High
- High
- Low
- Very Low
- "Wilderness"

The encroachment of roads in Finnmark, in northern Norway, between 1940 and 2000, and the associated loss of reindeer pasture.

Projected Infrastructure Development

2000 | 2030 | 2050

Projected development of infrastructure including roads, houses, and military training areas in northern Scandinavia 2000-2050.
The scenario illustrated here is based on the historical development of infrastructure, distribution and density of the human population, existing infrastructure, known location of oil, gas, mineral and forest resources, distance from coast, and vegetation type.

10 Multiple influences interact to cause impacts to people and ecosystems.

Human Health

Climate change will continue to affect human health in the Arctic. The impacts will differ from place to place due to regional differences in climate change as well as variations in health status and adaptive capacity of different populations. Rural arctic residents in small, isolated communities with a fragile system of support, little infrastructure, and marginal or non-existent public health systems appear to be most vulnerable. People who depend upon subsistence hunting and fishing, especially those who rely on just a few species, will be vulnerable to changes that heavily affect those species (for example, reduced sea ice and its impact on ringed seals and polar bears). Age, lifestyle, gender, access to resources, and other factors affect individual and collective adaptive capacity. And the historic ability to relocate to adapt to changing climatic conditions has been reduced as settlements have become permanent.

There are likely to be both adverse and beneficial outcomes of climate change on human health in the Arctic. Direct positive impacts could include a reduction in cold-induced injuries such as frostbite and hypothermia and a reduction in cold stress. Death rates are higher in winter than in summer and milder winters in some regions could reduce the number of deaths during winter months. However, the relationship between increased numbers of deaths and winter weather is difficult to interpret and more complex than the association between illness and death related to high temperatures. For example, many winter deaths are due to respiratory infections such as influenza, and it is unclear how higher winter temperatures would affect influenza transmission.

Direct negative impacts are likely to include increased heat stress and accidents associated with unusual ice and weather conditions. Indirect impacts include effects on diet due to changes in the access to and availability of subsistence foods, increased mental and social stresses related to changes in the environment and lifestyle, potential changes in bacterial and viral proliferation, mosquito-borne disease outbreaks, changes in access to good quality drinking water, and illnesses resulting from sanitation system problems. Health effects may also arise from interactions between contaminants, ultraviolet radiation, and climate change.

Indigenous people in some parts of the circumpolar North are reporting incidences of stress related to high temperature extremes not previously experienced. Impacts include respiratory difficulties, which, in turn, can limit an individual's participation in physical activities. However, fewer cold days associated with the warming trend in many regions during the winter are reported to have the positive effect of allowing people to get out more in the winter and alleviating stress related to extreme cold.

Climate-related changes in fish and wildlife distribution are very likely to result in significant changes in access to and the availability of traditional foods, with major health implications. A shift to a more Western diet is known to increase the risks of cancer, obesity, diabetes, and cardiovascular diseases among northern populations. Decreases in commercially important species, such as salmon, are likely to create economic hardship and health problems associated with reduced income in small communities.

Climate stress and shifting animal populations also create conditions for the spread of infectious diseases in animals that can be transmitted to humans, such as West Nile virus.

Safe drinking water and proper sanitation are critical to maintaining human health. Sanitation infrastructure includes water treatment and distribution systems, wastewater collection, treatment and disposal facilities, and solid waste collection and disposal. Permafrost thawing, coastal erosion and other climate-related changes that adversely affect drinking water quality, limit efficient delivery, or cause direct damage to facilities are likely to lead to adverse impacts on human health.

Increases in extreme events such as floods, storms, rockslides, and avalanches can be expected to cause an increase in injury and death. In addition to such direct impacts of these events, indirect effects could include impacts on the availability of safe drinking water. Intense rainfall events can also trigger mosquito-borne disease outbreaks, flood-related disasters, and, depending on existing water infrastructure, contamination of the water supply.

Mental health is also likely to be affected by climate related changes in the Arctic. Reduced opportunities for subsistence hunting, fishing, herding, and gathering are likely to cause psychological stresses due to the loss of important cultural activities. Flooding, erosion, and permafrost thawing related to climate change can negatively affect village habitability and infrastructure, and result in population dislocations and community disruption with resultant psychological impacts.

Rural arctic residents in small, isolated communities with a fragile system of support, little infrastructure, and marginal or non-existent public health systems appear to be most vulnerable.

West Nile Virus Change in Canada

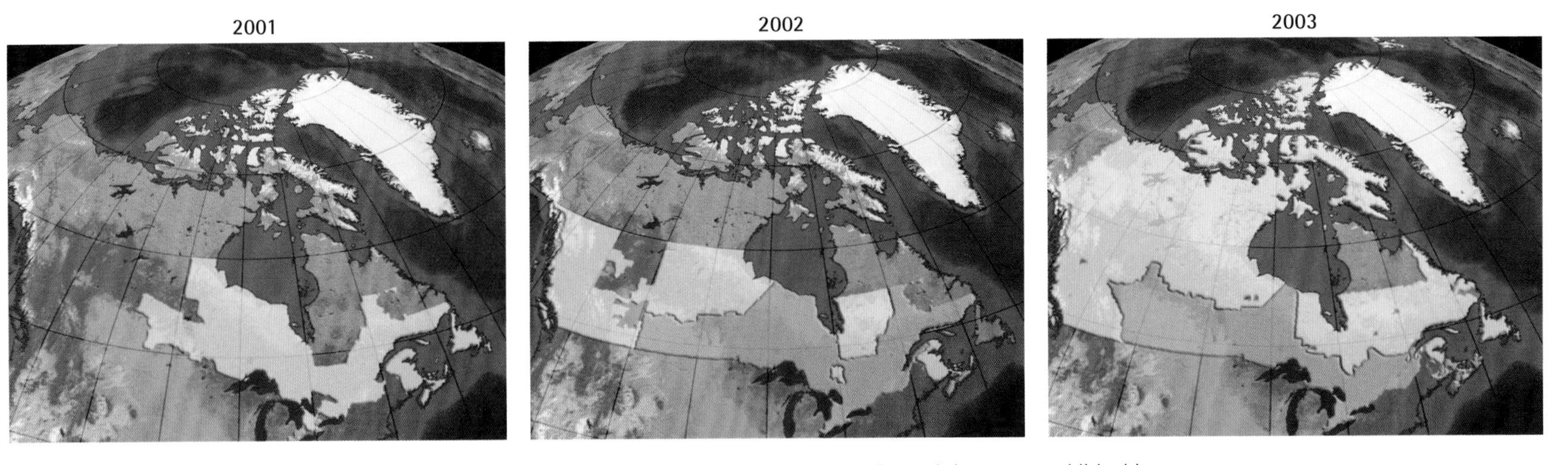

The West Nile encephalitis virus is a recent example of how far and fast a disease can spread once it becomes established in a new region. The West Nile virus can infect many bird and mammal species (including humans) and is transmitted by mosquitoes. It was first identified on the East Coast of North America in 1999 and spread to 43 states and six Canadian provinces by 2002. Migratory birds are responsible for its spread to other regions. Mosquitoes spread the virus to other birds (as well as to other animals and humans) within a region. Although the virus originated in tropical Africa, it has adapted to many North American mosquitoes, and so far, to over 110 species of North American bird, some of which migrate to the Arctic. Mosquito species known to transmit the virus are also found in the Arctic. Climate has historically limited the range of some insect-borne diseases, but climate change and adaptive disease agents such as the West Nile virus tend to favor continued northerly expansion. Some arctic regions, such as the State of Alaska, have initiated West Nile virus surveillance programs.

Dead Birds Submitted for Testing

Tested Positive for West Nile Virus

Selected Sub-Regional Impacts

About These Maps
The maps show observed and projected changes in climate in the four Arctic sub-regions as an annual average and for the winter season (December, January, and February).

The maps of observed temperature change show the difference from the middle of the 20th century to the present. For example, yellow indicates that an area has warmed by about 2°C over the past 50 years. Black indicates areas for which there are insufficient observational data to determine the extent of the change.

The maps for the future show the projected temperature change from the 1990s to the 2090s, based on the average change calculated by the five ACIA climate models using the lower of the two emissions scenarios (B2) considered in this assessment. On these maps, orange indicates that an area is projected to warm by about 6°C from the 1990s to the 2090s.

Changes in Climate in the ACIA Sub-Regions

Because the atmospheric and oceanic couplings to the rest of the world vary by sub-region, climate change has varied around the Arctic over the past century, with some sub-regions warming more than others and some even cooling slightly. Projections suggest that all parts of the Arctic will warm in the future, with some warming more than others.

Some of the sub-regional variations are likely to result from shifts in atmospheric circulation patterns. For example, Region I is particularly susceptible to changes in the North Atlantic Oscillation, which is a variation in the strength of the eastward airflow across the North Atlantic Ocean and into Europe. When the eastward airflow is strong, warm maritime air penetrates northern Eurasia and the Arctic during winter, resulting in warmer-than-normal conditions. This airflow pattern is consistent with and may be responsible for some of the warming of the Eurasian Arctic in recent decades. A critical issue in projections of 21st-century climate for this region is the state of the North Atlantic Oscillation, including its possible response to increasing greenhouse gas concentrations.

SUB-REGION I (East Greenland, Iceland, Norway, Sweden, Finland, Northwest Russia, and adjacent seas)

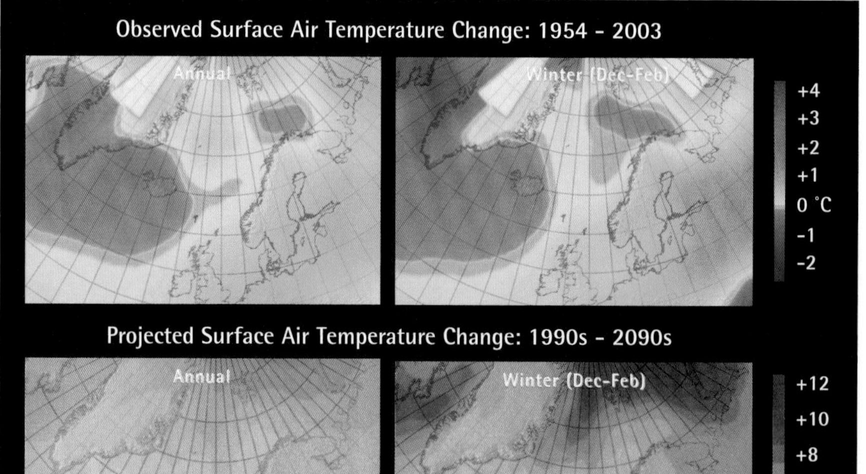

Over the last 50 years, annual average temperatures have increased by about 1°C over East Greenland, Scandinavia, and Northwest Russia, while there has been cooling of up to 1°C over Iceland and the North Atlantic Ocean. Near surface air temperatures over the Arctic and North Atlantic Oceans have remained very cold in winter, limiting the warming in coastal areas. Over inland areas, however, average wintertime temperatures have increased by about 2°C over Scandinavia and 2-3°C over Northwest Russia.

By the 2090s, model simulations project additional annual average warming of around 3°C for Scandinavia and East Greenland, about 2°C for Iceland, and roughly 6°C over the central Arctic Ocean. Average wintertime temperatures are projected to rise by 3-5°C over most land areas and up to 6°C over Northwest Russia, with the increase becoming larger near the coasts as a result of the 6-10°C warming over the nearby Arctic Ocean.

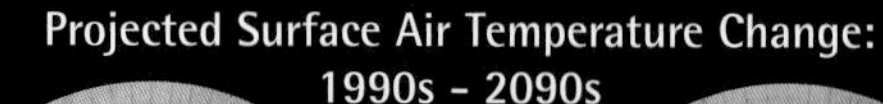

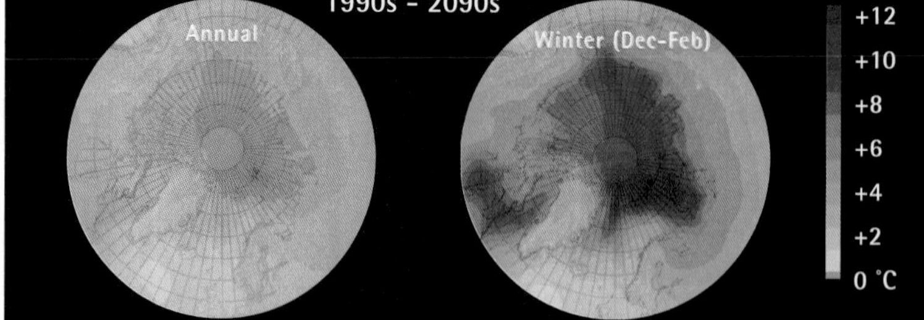

The Central Arctic Ocean is projected by all the models to warm more strongly than any of the four sub-regions, warming by up to 7°C annually and by up to 10°C in winter by the 2090s.

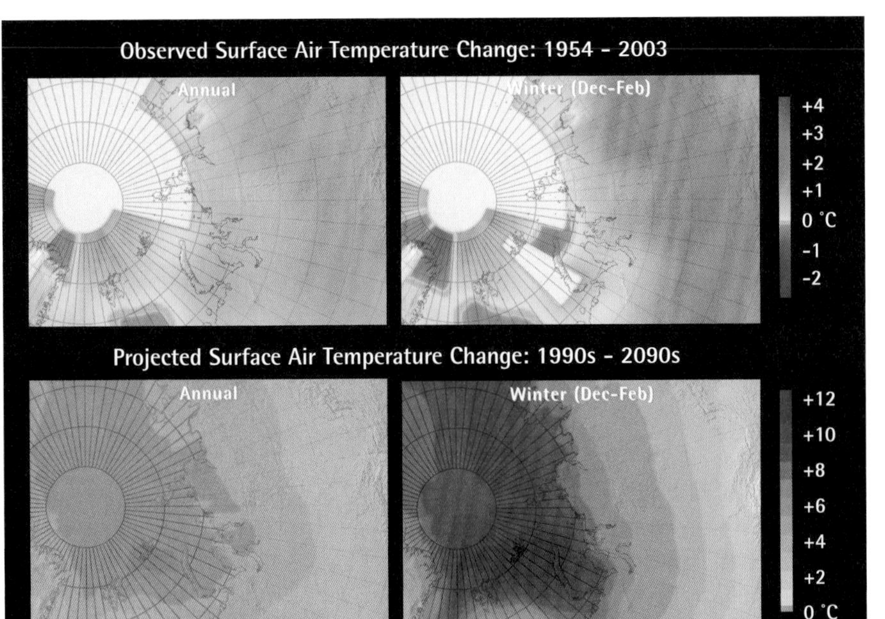

SUB-REGION II (Siberia and adjacent seas)

Annual average temperatures over Siberia have increased by about 1-3°C over the past 50 years, with most of the warming occurring during the winter, when temperatures increased about 3-5°C. The largest warming occurred inland in areas where reduced duration of snow cover helped amplify the warming.

By the 2090s, model simulations project additional annual average warming of around 3-5°C over land, with the increase becoming greater closer to the Arctic Ocean where air temperatures are expected to rise by about 5-7°C. Wintertime increases are projected to be 3-7°C over land, also increasing near Siberia's northern coastline due to the increases of 10°C or more over the adjacent ocean areas.

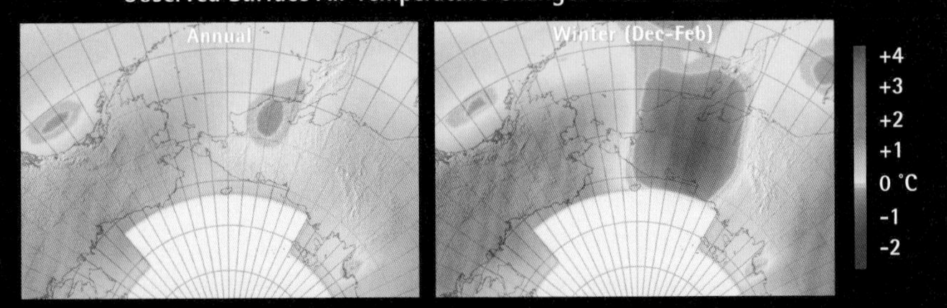

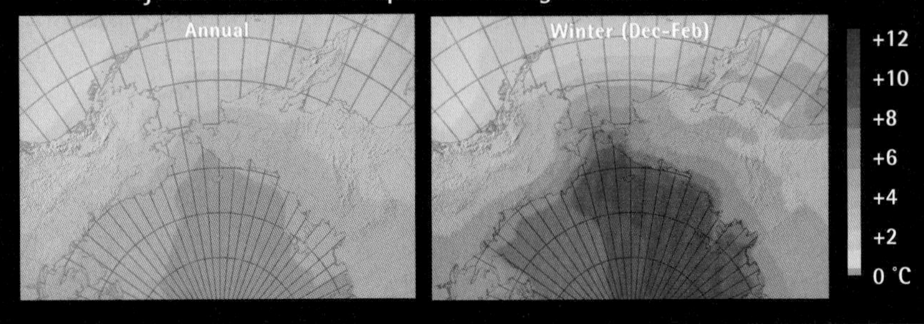

SUB-REGION III (Chukotka, Alaska, Western Canadian Arctic, and adjacent seas)

Over the last 50 years, annual average temperatures have risen by about 2-3°C in Alaska and the Canadian Yukon, and by about 0.5°C over the Bering Sea and most of Chukotka. The largest changes have been during winter, when near-surface air temperatures increased by about 3-5°C over Alaska, the Canadian Yukon, and the Bering Sea, while winters in Chukotka got 1-2°C colder.

For the 2090s, model simulations project annual average warming of 3-4°C over the land areas and Bering Sea, and about 6°C over the central Arctic Ocean. Winter temperatures are projected to rise by 4-7°C over the land areas, and up to 10°C over the Arctic Ocean.

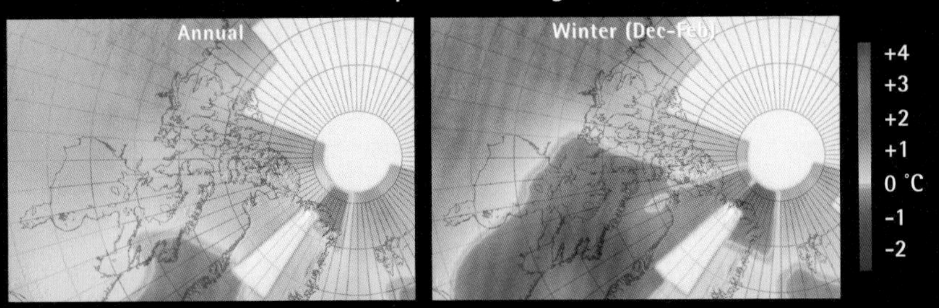

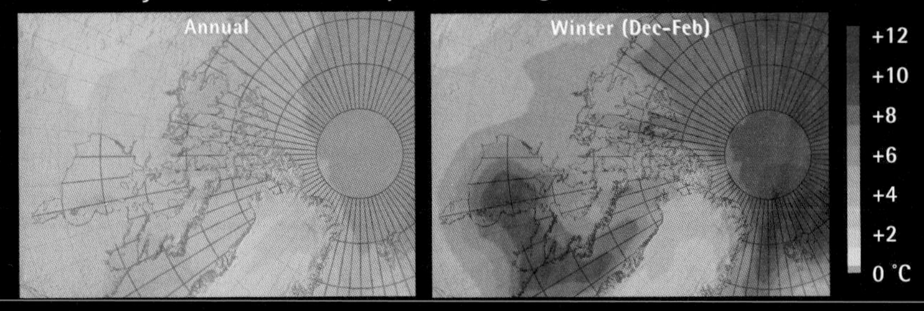

SUB-REGION IV (Central and Eastern Canadian Arctic, West Greenland, and adjacent seas)

Over the past 50 years, annual average temperatures increased by roughly 1-2°C over most of the Canadian Arctic and northwest Greenland. The Labrador Sea remained cold and nearby areas of Canada and southwest Greenland cooled by up to 1°C. Wintertime temperatures over central Canada increased by as much as 3-5°C, while areas of Canada and Greenland surrounding the Labrador Sea cooled by as much as 1-2°C.

By the 2090s, the entire region shows warming. Average annual warming of up to 3-5°C is projected over the Canadian Archipelago and 5-7°C over the oceans. Wintertime temperatures are projected to increase by 4-7°C over most of Canada and 3-5°C over Greenland, with increases of 8 to more than 10°C over Hudson Bay, the northern Labrador Sea, and the Arctic Ocean as sea ice declines.

Key Impacts - SUB-REGION I

East Greenland, Iceland, Norway, Sweden, Finland, Northwest Russia, and adjacent seas

Species Impacts Due to Sea-Ice Decline

Major decreases in sea-ice cover in summer and earlier ice melt and later freeze-up will have a variety of impacts in this region. As examples, the reduced reflectivity of the ocean's surface will increase regional and global warming; the reduction in sea ice is likely to enhance productivity at the base of the marine food chain, possibly increasing the productivity of some fisheries; sea-ice retreat will decrease habitat for polar bears and ice-living seals to an extent likely to threaten the survival of these species in this region; and more open water is likely to benefit some whale species.

Forest Changes

Observations in this region indicate that treelines advanced upslope by up to 60 meters in altitude in northern Sweden during the 20th century. The rate of advance in recent decades has been half a meter per year and 40 meters per °C. In the Russian part of this region, there has actually been a southward shift in treeline, apparently associated with pollution, deforestation, agriculture, and the growth of bogs that leads to the death of trees. In some areas of Finland and northern Sweden, an apparent increase in rapidly changing warm and cold episodes in winter has led to increasing bud damage in birch trees.

Projected warming is very likely to cause northward shifts of the boreal conifer forest and woodlands and the arctic/alpine tundra of this region. The potential for vegetation change is perhaps greatest in northern Scandinavia, where large shifts occurred historically in response to warming. In this area, the pine forest is expected to invade the lower belt of mountain birch forest, while the birch treeline is projected to move upward in altitude and northward, displacing shrub tundra vegetation, which would, in turn, displace alpine tundra. Warmer winters are expected to result in an increase in insect damage to forests. Some of the larger butterflies and moths have already been observed to be expanding their ranges northward, and some of their larvae are known to defoliate local tree species.

Biodiversity Loss

In this sub-region, recent warmer winters and changing snow conditions are thought to have contributed to declines in some reindeer populations and to the observed collapse in lemming and small rodent population peaks in recent decades. Such collapses in turn lead to a decrease in populations of birds and other animals, with the most severe declines in carnivores such as arctic foxes and raptors such as snowy owls. Populations of these two species are already in decline, along with several other bird species. As species ranges shift northward, alpine species in northern Norway, Sweden, Finland, and Russia are most threatened because there is nowhere for them to go as suitable habitats disappear from the mainland. The strip of tundra habitat between the forest and the ocean is particularly narrow and vulnerable to loss.

For freshwater fish species in this region, local diversity is projected to increase initially as new species migrate northward. However, as warming continues in the decades to come, temperatures are very likely to exceed the thermal tolerances of some native species, thus decreasing species diversity. The end result may be a similar number of species, but a different species composition, with some species added and others lost. However, in general, the species added to the Arctic will be those from lower latitudes, while those lost are very likely to be lost globally as there is nowhere else for them to go. The end result would be a global loss in biodiversity.

"The weather has changed to worse and to us it is a bad thing. It affects mobility at work. In the olden days the permanent ice cover came in October... These days you can venture to the ice only beginning in December. This is how things have changed."

Arkady Khodzinsky
Lovozero, Russia

...On the Economy

Marine Fisheries

This region is home to some of the most productive marine fishing grounds in the Arctic. Higher ocean temperatures are likely to cause northward shifts of some fish species, as well as changes in the timing of their migration, possible expansion of feeding areas, and increased growth rates. Under a moderate warming scenario, it is possible that a valuable cod stock could be established in West Greenland waters if larvae drifted over from Iceland and if fishing pressure were kept off long enough to allow a spawning stock to become established. On the other hand, under those circumstances, northern shrimp catches would be expected to decline by 70%, since these shrimp are an important part of the diet of cod. More southern fish species, such as mackerel could move into the region, providing a new opportunity, although capelin catches would be likely to dwindle.

Forestry

Forestry has already been affected by climate change and impacts are likely to become more severe in the future. Forest pest outbreaks in the Russian part of the region have caused the most extensive damage. The European pine sawfly affected a number of areas, each covering more than 5000 hectares. The annual number of insect outbreaks in 1989-1998 was 3.5 times higher than in 1956-1965 and the average intensity of forest damage doubled. While most of the region has seen modest growth in forestry, Russia has experienced a decline due to political and economic factors. These factors are likely to be aggravated by warming, which in the short term negatively impacts timber quality through insect damage, and infrastructure and winter transport through ground thawing.

...On People's Lives

Reindeer Herding

Reindeer herding by the Saami and other Indigenous Peoples is an important economic and cultural activity in this region, and people who herd reindeer are concerned about the impacts of climate change. In recent years, autumn weather in some areas has fluctuated between raining and freezing, often creating an ice layer on the ground that has reduced reindeer's access to the underlying lichen. These conditions represent a major change from the norm, and in some years, have resulted in extensive losses of reindeer. Changes in snow conditions also pose problems. When herding has become motorized, herders relying on snowmobiles have had to delay moving their herds until the first snows. In some years, this has led to delays up to mid-November. Also, the terrain has often been too difficult to travel over when the snow cover is light. Future changes in snow extent and condition have the potential to lead to major adverse consequences for reindeer herding and the associated physical, social, and cultural livelihood of the herders.

Socioeconomic Changes

The prospects and opportunities of gaining access to important natural resources have attracted a large number of people to this region. The relatively intense industrial activities, particularly on the Kola Peninsula, have resulted in population densities that are the highest in the circumpolar North. Increased opportunities for agriculture are projected as warming progresses. Impacts of climate change and their implications for the availability of resources could lead to major changes in economic conditions and subsequent shifts in demographics, societal structure, and cultural traditions of the region.

Key Impacts - SUB-REGION II

Siberia and adjacent seas

...On the Environment

Siberian River Flows

Changes in climate will have major impacts on the large Siberian rivers that flow into the Arctic. Projected increases in wintertime precipitation will increase river runoff, with a projected 15% annual increase in freshwater entering the Arctic Ocean by the later decades of this century, and a shift in the timing of peak flows to earlier in the spring. Greater winter and spring runoff will increase flows of nutrients and sediments to the Arctic Ocean, resulting in both positive and negative impacts. Coastal wetland and bog ecosystems are likely to expand, adding habitat for some species, but also increasing methane emissions. The projected increase in freshwater input to the ocean is likely to have important implications for factors that influence ocean currents and sea ice, with global as well as regional impacts. The increased water flows across the coastal zone are also likely to accelerate the thawing of coastal and sub-sea permafrost along most of the region's coastline.

Precipitation and Soils

The expected increase in precipitation will generally lead to wetter soils when soils are not frozen, and greater ice content of upper soil layers during winter. While snowfall during winter is likely to increase, the duration of the snow cover season is expected to shorten as warming accompanies the increased precipitation. The projected increase in moisture availability is likely to favor plant growth in areas that are otherwise moisture-limited.

...On the Economy

Northern Sea Route Opening

A potentially major impact on the region's economy could be the opening of the Northern Sea Route to commercial shipping. Summertime access to most coastal waters of the Eurasian Arctic is projected to be relatively ice-free within a few decades, with much more extensive melting later in the century. With the continued retreat of winter multi-year sea ice in the Arctic Ocean, it is plausible that the entire Eurasian maritime Arctic will be dominated by first-year sea ice in winter, with a decreasing frequency of multi-year ice intrusions into the coastal seas and more open water during summer. Such a change is likely to have important implications for route selection in this region. By the end of this century, the length of the navigation season (the period with sea ice concentrations below 50%) along the Northern Sea Route is projected to increase to about 120 days from the current 20-30 days.

Coal and Mineral Transport

The coal and mineral extraction industries are important parts of Russia's economy. Transportation of coal and minerals is likely to be affected in both positive and negative ways by climate change. Mines in Siberia that export their products by ocean shipping are very likely to experience savings due to reduced sea ice and a longer navigation season. Mining facilities that rely on roads over permafrost for transport are very likely to experience higher maintenance costs as permafrost thaws. The oil and natural gas industries are likely to be similarly affected, with improved access by sea and more problematic access on land.

The projected increase in freshwater input to the ocean is likely to have important implications for factors that influence ocean currents and sea ice, with global as well as regional impacts.

...On People's Lives

Water Resources

The change to a wetter climate is likely to lead to increased water resources for the region's residents. In permafrost-free areas, water tables are very likely to be closer to the surface, and more moisture is projected to be available for agricultural production. During the spring, when increased precipitation and runoff are very likely to cause higher river levels, the risk of flooding will increase. Lower water levels are projected for the summer, when they are likely to negatively affect river navigation and hydroelectric power generation and increase the risk of forest fires.

Infrastructure Damage

The combination of rising ground temperatures and inadequate design and construction practices for building on permafrost have resulted in major damage to infrastructure in Siberia in recent decades. Surveys in the 1990s in the region found nearly half of all buildings to be in poor condition, with buildings considered dangerous ranging from 22% in the village of Tiksi to 80% in the city of Vorkuta. In the last decade, building deformations increased to 42% in Norilsk, 61% in Yakutsk, and 90% in Amderma. Land transport routes are also faring poorly. In the early 1990s, 10-16% of sub-grade train tracks in the permafrost zone on the Baikal-Amur line were deformed because of permafrost thawing; this increased to 46% by 1998. The majority of airport runways in Norilsk, Yakutsk, Magadan, and other cities are currently in an emergency state. Damage to oil and gas transmission lines in the permafrost zone presents a particularly serious situation; 16 breaks were recorded on the Messoyakha-Norilsk pipeline in the last year. In the Khanty-Mansi autonomous district, 1702 accidents involving spills occurred and more than 640 square kilometers of land were removed from use in one year because of soil contamination.

Savings on Heating Cost

A reduction in the demand for heating fuel is a potential positive effect of climatic warming in this and other sub-regions. In Eastern Europe and Russia, most urban buildings have centralized heating systems that operate throughout the winter. Under scenarios of future warming, the duration of the period when building heating is required and the amount of energy required for heating are likely to decrease. The energy savings from decreased demand for heating in northern areas are likely to be offset by increases in the temperature and duration of the warm season in more southern parts of the region, where air conditioning will become desirable.

Impacts on Indigenous People

Many indigenous people of this region are reindeer herders. Large areas of pastureland are being lost to petroleum extraction and other industrial activities. Climate change is likely to add a new set of stresses. Frozen ground underlies most of the region and if warming degrades this permafrost, traditional reindeer migration routes are very likely to be disrupted. Warming is also projected to cause earlier melting and later freezing of sea ice in the Ob River delta, which could cut off access between winter and summer pastures. In addition, retreating sea ice will increase access to the region via the Northern Sea Route; this is likely to increase development, with potentially detrimental effects on local people and their traditional cultures.

KEY IMPACTS - SUB-REGION III

Chukotka, Alaska, Western Canadian Arctic and adjacent seas

...On the Environment

Forest Changes

This sub-region, especially Alaska and the Canadian Yukon, has experienced the most dramatic warming of all the sub-regions, resulting in major ecological impacts. Rising temperatures have caused northward expansion of boreal forest in some areas, significant increases in fire frequency and intensity, and unprecedented insect outbreaks; these trends are projected to increase. One projection suggests a threefold increase in the total area burned per decade, destroying coniferous forests and eventually leading to a deciduous forest-dominated landscape on the Seward Peninsula in Alaska, which is presently dominated by tundra. Some forested areas are likely to convert to bogs as permafrost thaws. The observed 20% increase in growing-degree days has benefited agriculture and forest productivity on some sites, while reducing growth on other sites.

Marine Species Impacts

Recent climate-related impacts observed in the Bering Sea include significant reductions in seabird and marine mammal populations, unusual algal blooms, abnormally high water temperatures, and low harvests of salmon on their return to spawning areas. While the Bering Sea fishery has become one of the world's largest, over the past few decades, the abundance of sea lions has declined between 50% and 80%. Numbers of northern fur seal pups on the Pribilof Islands – the major Bering Sea breeding grounds – declined by half between the 1950s and 1980s. There have been significant declines in the populations of some seabird species, including common murres, thick-billed murres, and red- and blacklegged kittiwakes. Numbers of salmon have been far below expected levels, fish have been smaller than average, and their traditional migratory patterns appear to have been altered. Future projections for the Bering Sea suggest productivity increases at the base of the food chain, poleward shifts of some cold-water species, and negative effects on ice-dwelling species.

Biodiversity at Risk

Arctic biodiversity is highly concentrated in this region, which is home to over 70% of the rare arctic plant species that occur nowhere else on earth. This region also contains significantly more threatened animal and plant species than any other arctic sub-region, making the biodiversity of this region quite vulnerable to climate change. Species concentrated in small areas, such as Wrangel Island, are particularly at risk from the direct effects of climate change coupled with the threat of non-native species that will move in and provide competition as climate warms. Northward expansion of dwarf shrub and tree dominated vegetation into Wrangel Island could result in the loss of many plant species. This region contains a long list of threatened species including the Wrangel lemming, whooping crane, Steller's sea eagle, lesser white-fronted goose, and the spoonbill sandpiper.

...On the Economy

Oil and Gas Industries

Extensive oil and gas reserves have been discovered in Alaska along the Beaufort Sea coast and in Canada's Mackenzie River/Beaufort Sea area. Climate impacts on oil and gas development in the region are likely to result in both financial benefits and costs in the future. For example, offshore oil exploration and production are likely to benefit from less extensive and thinner sea ice, although equipment will have to be designed to withstand increased wave forces and ice movement.

Ice roads, now used widely for access to facilities, are likely to be useable for shorter periods and to be less safe; this also applies to over-snow transport when there is less snow for a shorter duration. As a result of the warming since 1970, the number of days in which oil and gas exploration on the Alaskan tundra has been allowed under state standards has already fallen from 200 to 100 days per year. The standards, based on tundra hardness and snow conditions, are designed to limit damage to the tundra. The thawing of permafrost, on which buildings, pipelines, airfields, and coastal installations supporting oil and gas development are located, is very likely to adversely affect these structures and increase the cost of maintaining them.

"Our community has seen real and dramatic effects as a result of the warming that is occurring in the Arctic Ocean and the arctic environment. In the springtime, we are seeing the ice disappearing faster, which reduces our hunting time for walrus, seals, and whales."

Caleb Pungowiyi
Nome, Alaska

Fisheries

It is difficult to project impacts on the lucrative Bering Sea fisheries because numerous factors other than climate are involved, including fisheries policies, market demands and prices, and harvesting practices and technologies. Large northward shifts in fish and shellfish species are expected to accompany a warmer climate. Relocating fisheries infrastructure including fishing vessels, ports, and processing plants, may become necessary, entailing financial costs. Warmer waters are likely to lead to increased primary production in some areas, but a decline in coldwater species such as salmon and pollock.

...On People's Lives

Traditional Livelihoods

Livelihoods that sustain indigenous communities include hunting, trapping, gathering, and fishing. While making significant contributions to the diet and health of many indigenous populations, these activities also play large and important social and cultural roles. These livelihoods are already being threatened by multiple climate-related factors, including reduced or displaced populations of marine mammals, seabirds, and other wildlife, and reduction and thinning of sea ice, making hunting more difficult and dangerous. The Porcupine Caribou Herd is of particular importance to Indigenous Peoples in Alaska and Canada's Yukon and Northwest Territories, and climate-related impacts on this herd are already being observed.

Salmon and other fish that go up-river from the sea to spawn make up 60% of the wildlife resources that provide food for local users. Recent declines in these fish populations have thus directly affected the dietary and economic well-being of these people. Climate change is likely to have significant impacts on the availability of key food sources by shifting the range and abundance of salmon, herring, walrus, seals, whales, caribou, moose, and various species of seabird and waterfowl. The continued decline of summer sea ice is likely to push the populations of polar bears and ringed seals toward extinction in this century, with major implications for people who depend on these species.

Coastal Infrastructure Threatened

Increases in the frequency and ferocity of storm surges have triggered increased coastal erosion that is already threatening several villages along the coasts of the Bering and Beaufort Seas. The only available option is to plan for relocation of the villages, which will be very costly. Storm surges have also reduced the protection of coastal habitats provided by barrier islands and spits, which are highly vulnerable to erosion and wave destruction. Other climate-related impacts on village infrastructure are projected to continue to increase. Water and sanitation infrastructure is threatened in many places by thawing permafrost. Roads, buildings, pipelines, powerlines, and other infrastructure are also threatened by coastal erosion and degrading permafrost.

Key Impacts - SUB-REGION IV

Central and Eastern Canadian Arctic, West Greenland, and adjacent seas

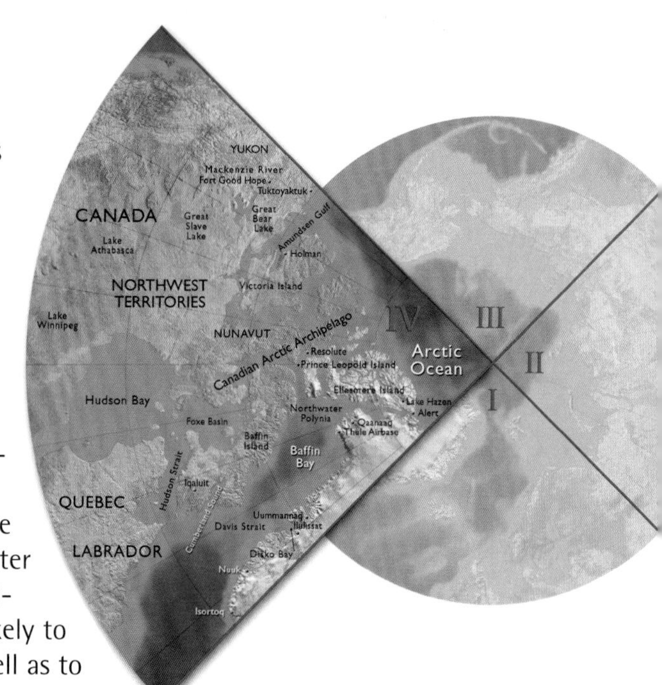

...On the Environment

Widespread Thawing

The maximum northward retreat of sea ice during the summer is projected to increase from the current 150-200 kilometers to 500-800 kilometers during this century. The thickness of fast ice (ice attached to the coast) in the Northwest Passage is projected to decrease substantially from its current one- to two-meter thickness. The Greenland Ice Sheet has experienced record melting in recent years and is likely to contribute substantially to sea-level rise as well as to possible changes in ocean circulation in the future. New research suggests that melting of the Greenland Ice Sheet is likely to occur more rapidly than previously believed.

Significant areas of permafrost in the Canadian part of this sub-region are at risk of thawing as air temperatures rise throughout this century. The boundary between continuous and discontinuous permafrost is projected to shift poleward by several hundred kilometers, resulting in the disappearance of a substantial amount of the permafrost in the present discontinuous zone. Many permafrost areas are also likely to experience more widespread thermokarsting (where the ground collapses due to thawing, producing craters or lakes) and increases in slope instability.

Ecosystem Shifts

Large ecosystem changes are projected. Shrinking of arctic tundra extent is very likely to result from a northward movement of treeline by as much as 750 kilometers in some areas. In recent decades, sparse stands of trees at the tundra edge in northeastern Canada have already begun filling in, creating dense stands that no longer retain the features of tundra. Forest health problems have become widespread in the region, driven by insects, fire, and tree stress all associated with recent mild winters and increasing heat during the growing season. It is very likely that such forest health problems will become increasingly intense and pervasive in response to future regional warming.

Changes in timing and abundance of forage availability, insect harassment, and parasite infestations will increase stress on caribou, tending to reduce their populations. North of the mainland, as the ability of High Arctic Peary caribou and musk ox to forage becomes increasingly limited as a result of adverse snow conditions, numbers will decline, with local extinctions in some areas. The fragmented land of the archipelago and large glaciated areas of the High Arctic in this sub-region constrain many land-based species from migrating as climate changes, placing them at greater risk than if they were on a mainland. In West Greenland, loss of habitat, displacement of species, and delayed migration of new species from the south will lead to a loss of present biodiversity.

If suitable pathways and habitats exist, ranges of many fish species in lakes and streams are likely to shift northward. Fish species in the southern part of the region such as Atlantic salmon and brook trout, are very likely to spread northward via near-shore marine waters, where they will out-compete more northerly local species such as Arctic char, causing local extinctions of these native species. Many marine mammal populations are likely to decline as sea ice recedes. The shortening of the sea ice season will negatively affect polar bear survival, decreasing populations, especially along southern margins of their distribution. Should the Arctic Ocean remain ice-free in summer for a number of consecutive years, it is likely that polar bears would be driven toward extinction.

...On the Economy

Increased Shipping

The costs and benefits of a longer shipping season in the Canadian Arctic areas are likely to be significant, but at this point, both are quite speculative. Increased ship traffic in the Northwest Passage, while providing economic opportunities, will also increase the risks and potential environmental damage from oil and other chemical spills. Increased costs are also likely to result from changes needed to cope with greater wave heights and possible flooding and erosion threats to coastal facilities. Increased sedimentation due to longer open water seasons could increase dredging costs.

Fisheries Changes

Under a moderate, gradual warming scenario, cod and capelin are likely to shift northward into the region, while northern shrimp and snow crabs are likely to decline. Many existing capelin-spawning beaches may disappear as sea level rises, potentially reducing survival. Seals are expected to experience higher pup mortality as sea ice thins and storm intensity increases. A reduction in the extent and duration of sea ice is likely to allow fishing further to the north, though it is also likely to reduce Greenland halibut fisheries that are conducted through fast ice.

In rivers and lakes, freshwater fish productivity is likely to increase initially as habitats warm and nutrient inputs increase. However, as critical thresholds are reached (such as thermal limits), arctic-adapted species are projected to decline; some of these fisheries are mainstays of local diets. Similarly, loss of suitable thermal habitat for fish such as lake trout will result in decreased growth and declines of many populations, with impacts on sport fisheries and local tourism.

Infrastructure Impacts

Use of ice roads in near-shore areas and over-snow transport on land, which are important at present, are already being impacted by a warming climate and are likely to be further curtailed in the future because of thawing ground, reduced snow cover, and shorter ice seasons. Higher air temperatures are likely to reduce the energy needed for heating buildings. The summer construction season is expected to lengthen. For the next 100 years at least, mostly negative impacts are projected for existing infrastructure such as northern pipelines, pile foundations in permafrost, bridges, pipeline river crossings, dikes, erosion protection structures, and stability of open pit mine walls.

...On People's Lives

Impacts on Indigenous Peoples

The health of indigenous people is likely to be affected through dietary, social, cultural, and other impacts of the projected changes in climate, many of which are already being observed. Climate change will affect the distribution and quality of animals and other resources on which the health and lifestyles of many northern communities are based. A shorter winter season, increased snowfall, and less extensive and thinner sea ice are likely to decrease opportunities for Indigenous Peoples to hunt and trap. Threats to the survival of polar bears and seals are of major concern in this sub-region.

Adapting to climate-driven changes is constrained by the present social and economic situations of Indigenous Peoples. For example, in the past, Inuit might have moved to follow animal movements. They now live in permanent settlements that foreclose this option. The impacts of climate change on Indigenous Peoples are also complicated by other factors such as resource regulations, industrial development, and global economic pressures. The potential for increased marine access to some of the region's resources through the Northwest Passage, while providing economic benefits to some, could pose problems for Indigenous Peoples in the region as the expansion of industrial activities can have cumulative effects on traditional lifestyles.

"Change has been so dramatic that during the coldest month of the year, the month of December 2001, torrential rains have fallen in the Thule region so much that there appeared a thick layer of solid ice on top of the sea ice and the surface of the land... which was very bad for the paws of our sled dogs."

Uusaqqak Qujaukitsoq
Qaanaaq, Greenland

Improving Future Assessments

The Arctic Climate Impact Assessment represents the first effort to comprehensively examine climate change and its impacts in the Arctic region. As such, it represents the beginning of a process. The assessment brought together the findings of hundreds of scientists from around the world whose research focuses on the Arctic. It also included the insights of Indigenous Peoples who have developed deep understandings through their long history of living and gathering knowledge in this region. Linking these scientific and indigenous perspectives is still in its early stages, and clearly has potential to improve understanding of climate change and its impacts. A great deal has been learned from the ACIA process and interactions, though much remains to be studied and better understood. This process should continue, with a focus on reducing uncertainties, filling gaps in knowledge identified during the assessment, and more explicitly including issues that interact with climate change and its impacts.

A critical self-assessment of the ACIA reveals achievements as well as deficiencies. The assessment covered potential arctic-wide impacts on the environment extensively. Estimates of economic impacts, on the other hand, and of impacts at the sub-regional level, were covered in a more cursory and exploratory manner, and greater development of such estimates must be a future priority task. Studies that integrate climate change impacts with effects due to other stresses (and thus assess the cumulative vulnerability of communities) were covered only in a preliminary fashion in this assessment.

Understanding and gaps in knowledge vary across the breadth of the assessment. Not all aspects need to be re-assessed comprehensively and not all aspects need to be assessed at the same time; some developments in science and some environmental changes take longer than others. Three major topics are thus suggested as future priorities for analysis: regional impacts, socioeconomic impacts, and vulnerabilities. These all involve improving the understanding of impacts on society. In each of these areas, involvement of a range of experts and stakeholders, especially including arctic indigenous communities, would help fill gaps in knowledge and provide relevant information to decision makers at all levels.

Sub-regional Impacts: There is a need to focus future assessments on smaller regions, perhaps at the local level, where an assessment of impacts of climate change has the greatest relevance and utility for residents and their activities.

Socioeconomic Impacts: Important economic sectors in the Arctic include oil and gas production, mining, transportation, fisheries, forestry, and tourism. Most of these sectors will experience direct and indirect impacts due to climate change, but in most cases, only qualitative information on economic impacts is presently available.

Assessing Vulnerabilities: Vulnerability is the degree to which a system is susceptible to adverse effects of multiple interacting stresses. Assessing vulnerability involves knowledge not just of the consequences of stresses and their interactions, but also of the capacity of the system to adapt.

To address these three high-priority research agendas will require a suite of improvements in long-term monitoring, process studies, climate modeling, and analyses of impacts on society.

Long-Term Monitoring: Long-term time series of climate and climate-related parameters are available from only a few locations in the Arctic. Continuation of long-term records is crucial, along with upgrading and expanding the observing systems that monitor snow and ice features, runoff from major rivers, ocean parameters, and changes in vegetation, biodiversity, and ecosystem processes.

Process Studies: Many arctic processes require further study, both through scientific investigations and through more detailed and systematic documentation of indigenous knowledge. Priorities include collection and interpretation of data related to climate and the physical environment, and studies of the rates and ranges of change for plants, animals, and ecosystem function. Such studies often involve linking climate models with models of ecosystem processes and other elements of the arctic system.

Modeling: Improvements in modeling arctic climate and its impacts are needed, including in the representation of ocean mixing and linkages to sea ice, permafrost-soil-vegetation interactions, important feedback processes, and extreme events. Model refinement and validation is required for models within scientific disciplines, and there is also a need to link and integrate models across disciplines. Developing, verifying, and applying very high-resolution coupled regional models to improve projections of regional changes in climate would also help provide more useful information to local decision-makers.

Analysis of Impacts on Society: Improving projections of the consequences of climate change on society will depend in part on the advances in climate modeling mentioned above as well as on generating improved scenarios of population and economic development in the Arctic, developing and applying impact scenarios, forging improved links between scientific and indigenous knowledge, and more thoroughly identifying and evaluating potential measures to mitigate and adapt to climate change.

Outreach in the Arctic

Finding effective ways of bringing the information gathered in the ACIA process to the communities of the Arctic presents an additional challenge. A variety of scientific, governmental, and non-governmental organizations plan to work to make the results of the ACIA process useful to a wide variety of constituents, from those who live and work on the land to those who determine local, national, and international policies relevant to the climate challenge.

International Linkages

The ACIA has built on the substance and conclusions of the assessments prepared by the Intergovernmental Panel on Climate Change (IPCC), which evaluate and summarize the world's most authoritative information regarding global climate change and its impacts. The most recent report of the IPCC, the Third Assessment Report, was released in 2001. The next IPCC assessment is in the early stages of development and is planned for publication in 2007. Just as the ACIA has built on IPCC's past evaluations, the 2007 IPCC report will build on ACIA's findings with regard to the Arctic, doing so in a way that adds more global context.

There are also other national and international efforts that offer opportunities to further understanding of the impacts of climate change and ultraviolet radiation. For example, the United Nations Environment Programme and the World Meteorological Organization have organized ongoing assessments of ozone depletion and its impacts. The International Conference on Arctic Research Planning II is drawing upon ACIA results as it develops a research agenda for the coming decades. The International Polar Year (IPY), being planned by the world's scientific community for 2007/9 will provide another opportunity to focus research attention on climate change and other important arctic issues. It was the International Geophysical Year in 1957/8 that initiated the first systematic measurements of stratospheric ozone and atmospheric carbon dioxide, thus laying the basis for the discoveries of ozone depletion and greenhouse gas-induced climate change. Without these decades of observations, the downward trend in stratospheric ozone and the continuous increase in atmospheric carbon dioxide could not have been detected.

The gaps in knowledge and needs for improved monitoring identified during the ACIA process are already affecting a variety of international research agendas. One of the primary goals already approved for the upcoming IPY is to study and evaluate present and future changes in climate in the polar regions and to evaluate the global-scale impacts of these changes. ACIA's findings can help to focus the research efforts of the IPY and other efforts. In turn, research initiated by other efforts can help fill the gaps that ACIA has identified in order to help carry out more detailed assessments of the importance of climate change for the Arctic.

Concluding Thoughts

As the scientific results presented in this assessment clearly illustrate, climate change presents a major and growing challenge to the Arctic and the world as a whole. While the concerns this generates are important now, their implications are of even greater importance for the future generations that will inherit the legacy of the current actions or inaction. Strong near-term action to reduce emissions is required in order to alter the future path of human-induced warming. Action is also needed to begin to adapt to the warming that is already occurring and will continue. The findings of this first Arctic Climate Impact Assessment provide a scientific basis upon which decision makers can consider, craft, and implement appropriate actions to respond to this important and far-reaching challenge.

Change Presents Risks and Opportunities

Trends are not destiny.

Rene Dubos

As this report has shown, climate change is very likely to result in major environmental changes that will present risks as well as opportunities across the Arctic. For example, the large reduction in summer sea ice threatens the future of several ice-dependent species including polar bears and seals, and thus the peoples that depend upon them. On the other hand, potential opportunities are likely to arise from expansion of marine access to resources, population centers, and distant markets via trans-arctic shipping routes.

Potential Surprises

Some of the climate-related changes in the arctic environment that are most likely to occur are expected to have major impacts; these include the decline in sea ice, the increase in coastal erosion, and the thawing of permafrost. In addition, other concerns emerge from possible outcomes that appear to have only a low likelihood, but the occurrence of which would have very large impacts – so-called "surprises". Due to the complexity of the Earth system, it is possible that climate change will evolve differently than the gradually changing scenarios used in this assessment. For example, storm intensities and tracks could change in unforeseen ways or temperatures could rise or fall abruptly due to unexpected disturbances of global weather systems. Possible changes in the global thermohaline circulation and widespread ramifications of such changes provide another example of a potential climate surprise. Although such changes could cause major impacts, very little information is currently available for considering such possibilities.

The Bottom Line

Despite the fact that a relatively small percentage of the world's greenhouse gas emissions originate in the Arctic, human-induced changes in arctic climate are among the largest on earth. As a consequence, the changes already underway in arctic landscapes, communities, and unique features provide an early indication for the rest of the world of the environmental and societal significance of global climate change. As this report illustrates, changes in climate and their impacts in the Arctic are already being widely noticed and felt, and are projected to become much greater. These changes will also reach far beyond the Arctic, affecting global climate, sea level, biodiversity, and many aspects of human social and economic systems. Climate change in the Arctic thus deserves and requires urgent attention by decision makers and the public worldwide.

Appendix - 1

The Emissions Scenarios Used in this Assessment

In its *Special Report on Emissions Scenarios* (SRES), the IPCC presented a wide range of plausible emissions scenarios for the 21st century based on various assumptions about future levels of population, economic growth, technological development, and other relevant factors. Of the six "illustrative scenarios" presented in the SRES, ACIA chose to focus primarily on one of these that fell slightly below the middle of the range of future emissions. That scenario, referred to as B2, is the basis for the projected climate maps in this report. A second scenario, A2, which falls above the middle of the SRES range, was also used in a few analyses, and is always identified as such. The focus on these scenarios here reflects a number of practical limits to conducting this assessment, and is not a judgment that these are the most likely outcomes.

Under all of the IPCC emissions scenarios, global carbon dioxide concentration, average surface air temperature, and sea level are projected to increase during the 21st century. From 2000 to 2100, the range of warming resulting from these scenarios is projected to be between 1.4 and 5.8°C. None of these scenarios include explicit policies to reduce greenhouse gas emissions. On the other hand, they do incorporate assumptions that involve major changes from the status quo for reasons other than limiting climate change, and these various factors influence the resulting levels of greenhouse gas emissions.

For example, the B2 emissions scenario assumes a world concerned with environmental protection and social equity, with solutions focused at the local and regional levels. It is a world in which global population grows to reach 10.4 billion by 2100, there is an intermediate level of economic development, and there is diverse technological change around the world. In a B2 world, by the year 2100, coal supplies 22% of the primary energy, and 49% of the world's energy is derived from sources that emit no carbon dioxide.

The A2 scenario also describes a world focused on self-reliance and preservation of local identities, but unlike B2, an A2 world is more concerned with economic growth than with environmental protection and social equity. Population growth is rapid, reaching 15 billion people by 2100. Economic development is primarily regionally oriented and per capita economic growth and technological change are relatively slow and fragmented. World GDP is slightly higher in 2100 in A2 than in B2. Coal provides 53% of the world's primary energy in 2100 in an A2 world, and 28% of the world's energy comes from sources than emit no carbon dioxide.

Other emission scenarios have been developed that consider the implication of policies that would reduce greenhouse gas emissions enough to stabilize their concentrations in the atmosphere at various levels, and thus limit the rate and magnitude of future climate change. Such scenarios were not considered in this assessment.

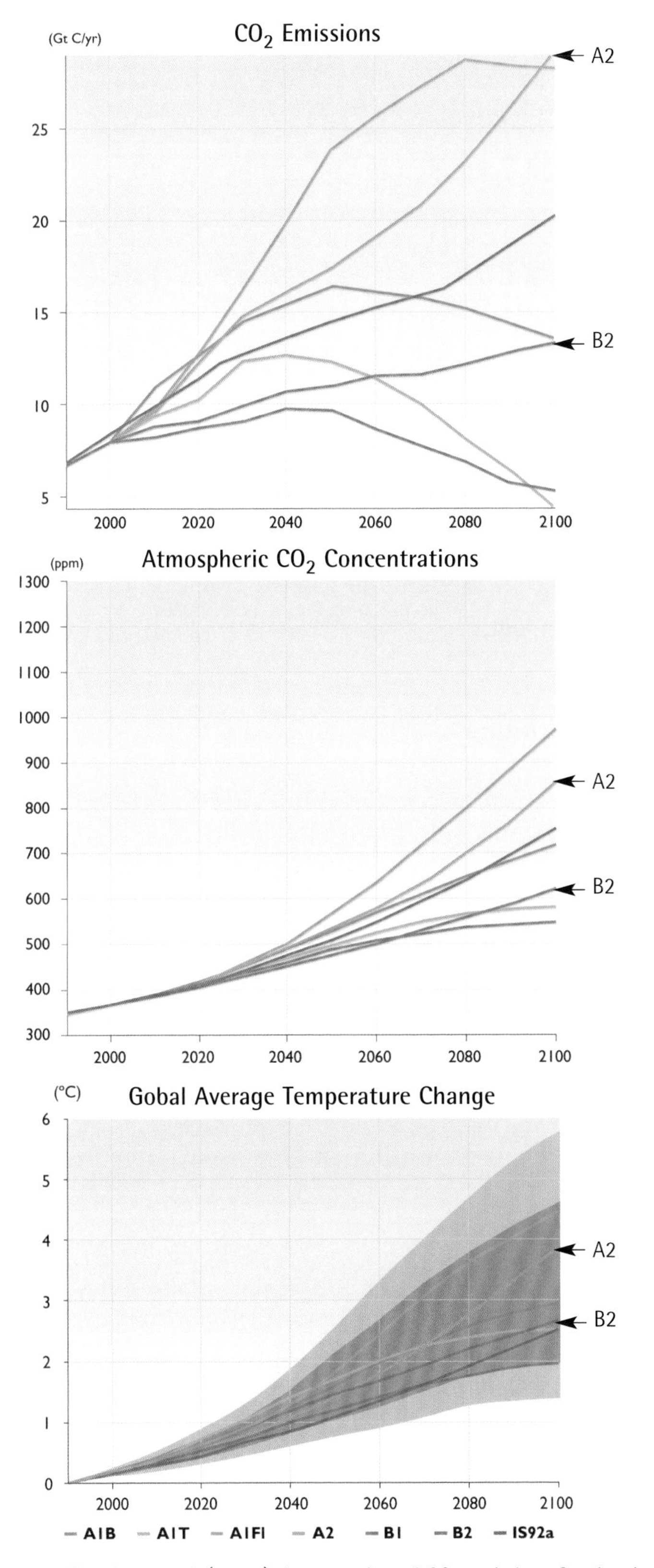

The first graph (upper) shows projected CO_2 emissions for the six illustrative IPCC SRES scenarios. The second graph (middle) shows the atmospheric CO_2 concentrations that would result from these emissions. The third graph (lower) shows the projected temperature trends that would result from these concentrations.

Primary Models used in the Arctic Climate Impact Assessment

CGCM2 - Canadian Centre for Climate Modelling and Analysis, Canada

CSM_1.4 - National Center for Atmospheric Research, United States

ECHAM4/OPYC3 - Max-Planck Institute for Meteorology, Germany

GFDL-R30_c - Geophysical Fluid Dynamics Laboratory, United States

HadCM3 - Hadley Centre for Climate Prediction and Research, United Kingdom

Five climate models from top research centers around the world were used in this assessment. Their names and the associated acronyms used throughout this report are listed above. These models all utilized the same emissions scenario, B2, described in this appendix. The climate maps shown throughout this report are based on these models, for the B2 emissions scenario.

Appendix - 2

Science Chapter Titles and Authors

Chapter 1: Introduction

Chapter 2: Arctic Climate – Past and Present

Chapter 3: The Changing Arctic: Indigenous Perspectives

Chapter 4: Future Climate Change: Modeling and Scenarios for the Arctic Region

Chapter 5: Ozone and Ultraviolet Radiation

Chapter 6: Cryospheric and Hydrologic Variability

Chapter 7: Arctic Tundra and Polar Desert Ecosystems

Chapter 8: Freshwater Ecosystems and Fisheries

Chapter 9: Marine Systems

Chapter 10: Principles of Conserving the Arctic's Biodiversity

Chapter 11: Management and Conservation of Wildlife in a Changing Arctic Environment

Chapter 12: Hunting, Herding, Fishing and Gathering: Indigenous Peoples and Renewable Resource Use in the Arctic

Chapter 13: Fisheries and Aquaculture

Chapter 14: Forests, Land Management and Agriculture

Chapter 15: Human Health

Chapter 16: Infrastructure: Buildings, Support Systems, and Industrial Facilities

Chapter 17: Climate Change in the Context of Multiple Stressors and Resilience

Chapter 18: Summary and Synthesis

Chapter 1: Introduction

Lead Author
Henry Huntington, Huntington Consulting, USA

Contributing Authors
Elizabeth Bush, Environment Canada, Canada
Terry V. Callaghan, Abisko Scientific Research Station, Sweden; Sheffield Centre for Arctic Ecology, UK
Vladimir M. Kattsov, Voeikov Main Geophysical Observatory, Russia
Mark Nuttall, University of Aberdeen, Scotland, UK; University of Alberta, Canada

Chapter 2: Arctic Climate – Past and Present

Lead Author
Gordon McBean, University of Western Ontario, Canada

Contributing Authors
Genrikh Alekseev, Arctic and Antarctic Research Institute, Russia
Deliang Chen, Göteborg University, Sweden
Eirik Førland, Norwegian Meteorological Institute, Norway
John Fyfe, Meteorological Service of Canada, Canada
Pavel Y. Groisman, NOAA National Climatic Data Center, USA
Roger King, The University of Western Ontario, Canada
Humfrey Melling, Fisheries and Oceans Canada, Canada
Russell Vose, NOAA National Climatic Data Center, USA
Paul H. Whitfield, Meteorological Service of Canada, Canada

Chapter 3: The Changing Arctic: Indigenous Perspectives

Lead Authors
Henry Huntington, Huntington Consulting, USA
Shari Fox, University of Colorado at Boulder, USA

Contributing Authors
Fikret Berkes, University of Manitoba, Canada
Igor Krupnik, Smithsonian Institution, USA

Case Study Authors
Kotzebue:
Alex Whiting, Native Village of Kotzebue, USA
The Aleutian and Pribilof Islands Region, Alaska:
Michael Zacharof, Aleutian International Association, USA
Greg McGlashan, St. George Tribal Ecosystem Office, USA
Michael Brubaker, Aleutian/Pribilof Islands Association, USA
Victoria Gofman, Aleut International Association, USA
The Yukon Territory:
Cindy Dickson, Arctic Athabascan Council, Canada
Denendeh:
Chris Paci, Arctic Athabaskan Council, Canada
Shirley Tsetta, Yellowknives Dene (N'dilo), Canada
Sam Gargan, Deh Gah Got'ine (Fort Providence), Canada
Chief Roy Fabian, Katloodeeche (Hay River Dene Reserve), Canada
Chief Jerry Paulette, Smith Landing First Nation, Canada
Vice-Chief Micheal Cazon, Deh Cho First Nations, Canada
Diane Giroux, former Sub-Chief Deninu K-ue (Fort Resolution), Canada
Pete King, Elder Akaitcho Territory, Canada
Maurice Boucher, Deninu K-ue (Fort Resolution), Canada
Louie Able, Elder Akaitcho Territory, Canada
Jean Norin, Elder Akaitcho Territory, Canada
Agatha Laboucan, Lutsel'Ke, Canada

Philip Cheezie, Elder Akaitcho Territory, Canada
Joseph Poitras, Elder, Canada
Flora Abraham, Elder, Canada
Bella T'selie, Sahtu Dene Council, Canada
Jim Pierrot, Elder Sahtu, Canada
Paul Cotchilly, Elder Sahtu, Canada
George Lafferty, Tlicho Government, Canada
James Rabesca, Tlicho Government, Canada
Eddie Camille, Elder Tlicho, Canada
John Edwards, Gwich'in Tribal Council, Canada
John Carmicheal, Elder Gwich'in, Canada
Woody Elias, Elder Gwich'in, Canada
Alison de Palham, Deh Cho First Nations, Canada
Laura Pitkanen, Deh Cho First Nations, Canada
Leo Norwegian, Elder Deh Cho, Canada
Nunavut:
Shari Fox, University of Colorado at Boulder, USA
Qaanaaq, Greenland:
Uusaqqak Qujaukitsoq, Inuit Circumpolar Conference, Greenland
Nuka Møller, Inuit Circumpolar Conference, Greenland
Saami:
Tero Mustonen, Tampere Polytechnic / Snowchange Project, Finland
Mika Nieminen, Tampere Polytechnic / Snowchange Project, Finland
Hanna Eklund, Tampere Polytechnic / Snowchange Project, Finland
Climate Change and the Saami:
Elina Helander, University of Lapland, Finland
Kola:
Tero Mustonen, Tampere Polytechnic / Snowchange Project, Finland
Sergey Zavalko, Murmansk State Technical University, Russia
Jyrki Terva, Tampere Polytechnic / Snowchange Project, Finland
Alexey Cherenkov, Murmansk State Technical University, Russia

Consulting Authors
Anne Henshaw, Bowdoin College, USA
Terry Fenge, Inuit Circumpolar Conference, Canada
Scot Nickels, Inuit Tapiriit Kanatami, Canada
Simon Wilson, Arctic Monitoring and Assessment Programme, Norway

Chapter 4: Future Climate Change: Modeling and Scenarios for the Arctic Region

Lead Authors
Erland Källén, Stockholm University, Sweden
Vladimir M. Kattsov, Voeikov Main Geophysical Observatory, Russia

Contributing Authors
Howard Cattle, International CLIVAR Project Office, UK
Jens Christensen, Danish Meteorological Institute, Denmark
Helge Drange, Nansen Environmental and Remote Sensing Center and Bjerknes Centre for Climate Research, Norway
Inger Hanssen-Bauer, Norwegian Meteorological Institute, Norway
Tómas Jóhannesen, Icelandic Meteorological Office, Iceland
Igor Karol, Voeikov Main Geophysical Observatory, Russia
Jouni Räisänen, University of Helsinki, Finland
Gunilla Svensson, Stockholm University, Sweden
Stanislav Vavulin, Voeikov Main Geophysical Observatory, Russia

Consulting Authors
Deliang Chen, Gothenburg University, Sweden
Igor Polyakov, University of Alaska Fairbanks, USA
Annette Rinke, Alfred Wegener Institute for Polar and Marine Research, Germany

Chapter 5: Ozone and Ultraviolet Radiation

Lead Authors
Betsy Weatherhead, University of Colorado at Boulder, USA
Aapo Tanskanen, Finnish Meteorological Institute, Finland
Amy Stevermer, University of Colorado at Boulder, USA

Contributing Authors
Signe Bech Andersen, Danish Meteorological Institute, Denmark
Antti Arola, Finnish Meteorological Institute, Finland
John Austin, University Corporation for Atmospheric Research/Geophysical Fluid Dynamics Laboratory, USA
Germar Bernhard, Biospherical Instruments Inc., USA
Howard Browman, Institute of Marine Research, Norway
Vitali Fioletov, Meteorological Service of Canada, Canada
Volker Grewe, DLR-Institut für Physik der Atmosphäre, Germany
Jay Herman, NASA Goddard Space Flight Center, USA
Weine Josefsson, Swedish Meteorological and Hydrological Institute, Sweden
Arve Kylling, Norwegian Institute for Air Research, Norway
Esko Kyro, Finnish Meteorological Institute, Finland
Anders Lindfors, Uppsala Astronomical Observatory, Sweden
Drew Shindell, NASA Goddard Institute for Space Studies, USA
Petteri Taalas, Finnish Meteorological Institute, Finland
David Tarasick, Meteorological Service of Canada, Canada

Consulting Authors
Valery Dorokhov, Central Aerological Observatory, Russia
Bjorn Johnsen, Norwegian Radiation Protection Authority, Norway
Jussi Kaurola, Finnish Meteorological Institute, Finland
Rigel Kivi, Finnish Meteorological Institute, Finland
Nikolay Krotkov, NASA Goddard Space Flight Center, USA
Kaisa Lakkala, Finnish Meteorological Institute, Finland
Jacqueline Lenoble, Université des Sciences et Technologies de Lille, France
David Sliney, U.S. Army Center for Health Promotion and Preventive Medicine, USA

Chapter 6: Cryospheric and Hydrologic Variability

Lead Author
John E. Walsh, University of Alaska Fairbanks, USA

Contributing Authors
Oleg Anisimov, State Hydrological Institute, Russia
Jon Ove M. Hagen, University of Oslo, Norway
Thor Jakobsson, Icelandic Meteorological Office, Iceland
Johannes Oerlemans, University of Utrecht, Netherlands
Terry Prowse, University of Victoria, Canada
Vladimir Romanovsky, University of Alaska Fairbanks, USA
Nina Savelieva, Pacific Oceanological Institute, Russia
Mark Serreze, University of Colorado at Boulder, USA
Alex Shiklomanov, University of New Hampshire, USA
Igor Shiklomanov, State Hydrological Institute, Russia
Steven Solomon, Geological Survey of Canada, Canada

Consulting Authors
Anthony Arendt, University of Alaska Fairbanks, USA
Michael N. Demuth, Natural Resources Canada, Canada
Julian Dowdeswell, Scott Polar Research Institute, UK
Mark Dyurgerov, University of Colorado at Boulder, USA
Andrey Glazovsky, Institute of Geography, RAS, Russia

Roy M. Koerner, Geological Survey of Canada, Canada
Niels Reeh, Technical University of Denmark, Denmark
Oddur Siggurdsson, National Energy Authority, Hydrological Service, Iceland
Konrad Steffen, University of Colorado at Boulder, USA
Martin Truffer, University of Alaska Fairbanks, USA

Chapter 7: Arctic Tundra and Polar Desert Ecosystems

Lead Author
Terry V. Callaghan, Abisko Scientific Research Station, Sweden; Sheffield Centre for Arctic Ecology, UK

Contributing Authors
Lars Olof Björn, Lund University, Sweden
F. Stuart Chapin III, University of Alaska Fairbanks, USA
Yuri Chernov, A.N. Severtsov Institute of Evolutionary Morphology and Animal Ecology, RAS, Russia
Torben R. Christensen, Lund University, Sweden
Brian Huntley, University of Durham, UK
Rolf Ims, University of Tromsø, Norway
Margareta Johansson, Abisko Scientific Research Station, Sweden
Dyanna Jolly Riedlinger, Dyanna Jolly Consulting, New Zealand
Sven Jonasson, University of Copenhagen, Denmark
Nadya Matveyeva, Komarov Botanical Institute, RAS, Russia
Walter Oechel, San Diego State University, USA
Nicolai Panikov, Stevens Technical University, USA
Gus Shaver, Marine Biological Laboratory, USA

Consulting Authors
Josef Elster, University of South Bohemia, Czech Republic
Heikki Henttonen, Finnish Forest Research Institute, Finland
Ingibjörg S. Jónsdóttir, University of Svalbard, Norway
Kari Laine, University of Oulu, Finland
Sibyll Schaphoff, Potsdam Institute for Climate Impact Research, Germany
Stephen Sitch, Potsdam Institute for Climate Impact Research, Germany
Erja Taulavuori, University of Oulu, Finland
Kari Taulavuori, University of Oulu, Finland
Christoph Zöckler, UNEP World Conservation Monitoring Centre, UK

Chapter 8: Freshwater Ecosystems and Fisheries

Lead Authors
Fred J. Wrona, National Water Research Institute, Canada
Terry D. Prowse, National Water Research Institute, Canada
James D. Reist, Fisheries and Oceans Canada, Canada

Contributing Authors
Richard Beamish, Fisheries and Oceans Canada, Canada
John J. Gibson, National Water Research Institute, Canada
John Hobbie, Marine Biological Laboratory, USA
Erik Jeppesen, National Environmental Research Institute, Denmark
Jackie King, Fisheries and Oceans Canada, Canada
Guenter Koeck, University of Innsbruck, Austria
Atte Korhola, University of Helsinki, Finland
Lucie Lévesque, National Water Research Institute, Canada
Rob Macdonald, Fisheries and Oceans Canada, Canada
Michael Power, University of Waterloo, Canada
Vladimir Skvortsov, Institute of Limnology, Russia
Warwick Vincent, Laval University, Canada

Consulting Authors
Robert Clark, Canadian Wildlife Service, Canada
Brian Dempson, Fisheries and Oceans Canada, Canada
David Lean, University of Ottawa, Canada
Hannu Lehtonen, University of Helsinki, Finland
Sofia Perin, University of Ottawa, Canada
Richard Pienitz, Laval University, Canada
Milla Rautio, Laval University, Canada
John Smol, Queen's University, Canada
Ross Tallman, Fisheries and Oceans Canada, Canada
Alexander Zhulidov, Centre for Preparation and Implementation of International Projects on Technical Assistance, Russia

Chapter 9: Marine Systems

Lead Author
Harald Loeng, Institute of Marine Research, Norway

Contributing Authors
Keith Brander, International Council for the Exploration of the Sea, Denmark
Eddy Carmack, Institute of Ocean Sciences, Canada
Stanislav Denisenko, Zoological Institute, RAS, Russia
Ken Drinkwater, Bedford Institute of Oceanography, Canada
Bogi Hansen, The Fisheries Laboratory, Faroe Islands
Kit Kovacs, Norwegian Polar Institute, Norway
Pat Livingston, NOAA National Marine Fisheries Service, USA
Fiona McLaughlin, Institute of Ocean Sciences, Canada
Egil Sakshaug, Norwegian University of Science and Technology, Norway

Consulting Authors
Richard Bellerby, Bjerknes Centre for Climate Research, Norway
Howard Browman, Institute of Marine Research, Norway
Tore Furevik, University of Bergen, Norway
Jacqueline M. Grebmeier, University of Tennessee, USA
Eystein Jansen, Bjerknes Centre for Climate Research, Norway
Steingrimur Jónsson, Marine Research Institute, Iceland
Lis Lindal Jørgensen, Institute of Marine Research, Norway
Svend-Aage Malmberg, Marine Research Institute, Iceland
Svein Østerhus, Bjerknes Centre for Climate Research, Norway
Geir Ottersen, Institute of Marine Research, Norway
Koji Shimada, Japan Marine Science and Technology Center, Japan

Chapter 10: Principles of Conserving the Arctic's Biodiversity

Lead Author
Michael B. Usher, University of Stirling, Scotland, UK

Contributing Authors
Terry V. Callaghan, Abisko Scientific Research Station, Sweden; Sheffield Centre for Arctic Ecology, UK
Grant Gilchrist, Canadian Wildlife Service, Canada
O.W. Heal, Durham University, UK
Glenn P. Juday, University of Alaska Fairbanks, USA
Harald Loeng, Institute of Marine Research, Norway
Magdalena A. K. Muir, Conservation of Arctic Flora and Fauna, Iceland
Pål Prestrud, Centre for Climate Research in Oslo, Norway

Chapter 11: Management and Conservation of Wildlife in a Changing Arctic Environment

Lead Author
David R. Klein, University of Alaska Fairbanks, USA

Contributing Authors
Leonid M. Baskin, Institute of Ecology and Evolution, Russia
Lyudmila S. Bogoslovskaya, Russian Institute of Cultural and Natural Heritage, Russia
Kjell Danell, Swedish University of Agricultural Sciences, Sweden
Anne Gunn, Government of the Northwest Territory, Canada
David B. Irons, U.S. Fish and Wildlife Service, USA
Gary P. Kofinas, University of Alaska Fairbanks, USA
Kit M. Kovacs, Norwegian Polar Institute, Norway
Margarita Magomedova, Institute of Plant and Animal Ecology, Russia
Rosa H. Meehan, U.S. Fish and Wildlife Service, USA
Don E. Russell, Canadian Wildlife Service, Canada
Patrick Valkenburg, Alaska Department of Fish and Game, USA

Chapter 12: Hunting, Herding, Fishing and Gathering: Indigenous Peoples and Renewable Resource Use in the Arctic

Lead Author
Mark Nuttall, University of Aberdeen, Scotland, UK; University of Alberta, Canada

Contributing Authors
Fikret Berkes, University of Manitoba, Canada
Bruce Forbes, University of Lapland, Finland
Gary Kofinas, University of Alaska Fairbanks, USA
Tatiana Vlassova, Russian Association of Indigenous Peoples of the North (RAIPON), Russia
George Wenzel, McGill University, Canada

Chapter 13: Fisheries and Aquaculture

Lead Authors
Hjalmar Vilhjalmsson, Marine Research Institute, Iceland
Alf Håkon Hoel, University of Tromsø, Norway

Contributing Authors
Sveinn Agnarsson, University of Iceland, Iceland
Ragnar Arnason, University of Iceland, Iceland
James E. Carscadden, Fisheries and Oceans Canada, Canada
Arne Eide, University of Tromsø, Norway
David Fluharty, University of Washington, USA
Geir Hønneland, Fridtjof Nansen Institute, Norway
Carsten Hvingel, Greenland Institute of Natural Science, Greenland
Jakob Jakobsson, Marine Research Institute, Iceland
George Lilly, Fisheries and Oceans Canada, Canada
Odd Nakken, Institute of Marine Research, Norway
Vladimir Radchenko, Sakhalin Research Institute of Fisheries and Oceanography, Russia
Susanne Ramstad, Norwegian Polar Institute, Norway
William Schrank, Memorial University of Newfoundland, Canada
Niels Vestergaard, University of Southern Denmark, Denmark
Thomas Wilderbuer, NOAA National Marine Fisheries Service, USA

Chapter 14: Forests, Land Management and Agriculture

Lead Author
Glenn P. Juday, University of Alaska Fairbanks, USA

Contributing Authors
Valerie Barber, University of Alaska Fairbanks, USA
Hans Linderholm, Göteborg University, Sweden
Scott Rupp, University of Alaska Fairbanks, USA
Steve Sparrow, University of Alaska Fairbanks, USA
Eugene Vaganov, V.N. Sukachev Institute of Forest Research, RAS, Russia
John Yarie, University of Alaska Fairbanks, USA

Consulting Authors
Edward Berg, U.S. Fish and Wildlife Service, USA
Rosanne D'Arrigo, Lamont Doherty Earth Observatory, USA
Paul Duffy, University of Alaska Fairbanks, USA
Olafur Eggertsson, Icelandic Forest Research, Iceland
V.V. Furyaev, V.N. Sukachev Institute of Forest Research, RAS, Russia
Edward H. (Ted) Hogg, Canadian Forest Service, Canada
Satu Huttunen, University of Oulu, Finland
Gordon Jacoby, Lamont Doherty Earth Observatory, USA
V. Ya. Kaplunov, V.N. Sukachev Institute of Forest Research, RAS, Russia
Seppo Kellomaki, University of Joensuu, Finland
A.V. Kirdyanov, V.N. Sukachev Institute of Forest Research, RAS, Russia
Carol E. Lewis, University of Alaska Fairbanks, USA
Sune Linder, Swedish University of Agricultural Sciences, Sweden
M.M. Naurzbaev, V.N. Sukachev Institute of Forest Research, RAS, USA
F.I. Pleshikov, V.N. Sukachev Institute of Forest Research, RAS, Russia
Ulf T. Runesson, Lakehead University, Canada
Yu.V. Savva, V.N. Sukachev Institute of Forest Research, RAS, Rusia
O.V. Sidorova, V.N. Sukachev Institute of Forest Research, RAS, Russia
V.D. Stakanov, V.N. Sukachev Institute of Forest Research, RAS, Russia
N.M. Tchebakova, V.N. Sukachev Institute of Forest Research, RAS, Russia
E.N. Valendik, V.N. Sukachev Institute of Forest Research, RAS, Russia
E.F. Vedrova, V.N. Sukachev Institute of Forest Research, RAS, Russia
Martin Wilmking, Lamont Doherty Earth Observatory, USA

Chapter 15: Human Health

Lead Authors
Jim Berner, Alaska Native Tribal Health Consortium, USA
Christopher Furgal, Laval University, Canada

Contributing Authors:
Peter Bjerregaard, National Institute of Public Health, Denmark
Mike Bradley, Alaska Native Tribal Health Consortium, USA
Tine Curtis, National Institute of Public Health, Denmark
Ed De Fabo, The George Washington University, USA
Juhani Hassi, University of Oulu, Finland
William Keatinge, Queen Mary and Westfield College, UK
Siv Kvernmo, University of Tromsø, Norway
Simo Nayha, University of Oulu, Finland
Hannu Rintamaki, Finnish Institute of Occupational Health, Finland
John Warren, Alaska Native Tribal Health Consortium, USA

Chapter 16: Infrastructure: Buildings, Support Systems, and Industrial Facilities

Lead Author
Arne Instanes, Instanes Consulting Engineers, Norway

Contributing Authors
Oleg Anisimov, State Hydrological Institute, Russia
Lawson Brigham, U.S. Arctic Research Commission, USA
Douglas Goering, University of Alaska Fairbanks, USA
Branko Ladanyi, École Polytechnique de Montreal, Canada
Jan Otto Larsen, Norwegian University of Science and Technology, Norway
Lev N. Khrustalev, Moscow State University, Russia

Consulting Authors
Orson Smith, University of Alaska Anchorage, USA
Amy Stevermer, University of Colorado at Boulder, USA
Betsy Weatherhead, University of Colorado at Boulder, USA
Gunter Weller, University of Alaska Fairbanks, USA

Chapter 17: Climate Change in the Context of Multiple Stressors and Resilience

Lead Authors
James J. McCarthy, Harvard University, USA
Marybeth Long Martello, Harvard University, USA

Contributing Authors
Robert Corell, American Meteorological Society and Harvard University, USA
Noelle Eckley, Harvard University, USA
Shari Fox, University of Colorado at Boulder, USA
Grete Hovelsrud-Broda, Centre for International Climate and Environmental Research, Norway
Svein Mathiesen, The Norwegian School of Veterinary Science and Nordic Sámi Institute, Norway
Colin Polsky, Clark University, USA
Henrik Selin, Boston University, USA
Nicholas Tyler, University of Tromsø, Norway

Consulting Authors
Kirsti Strøm Bull, University of Oslo and Nordic Sámi Institute, Norway
Inger Maria Gaup Eira, Nordic Sámi Institute, Norway
Nils Isak Eira, Fossbakken, Norway
Siri Eriksen, Centre for International Climate and Environmental Research, Norway
Inger Hanssen-Bauer, Norwegian Meteorological Institute, Norway
Johan Klemet Kalstad, Nordic Sámi Institute, Norway
Christian Nellemann, Norwegian Nature Research Institute, Norway
Nils Oskal, Sámi University College, Norway
Erik S. Reinert, Hvasser, Tønsberg, Norway
Douglas Siegel-Causey, Harvard University, USA
Paal Vegar Storeheier, University of Tromsø , Norway
Johan Mathis Turi, Association of World Reindeer Herders, Norway

Chapter 18: Summary and Synthesis

Lead Author
Gunter Weller, University of Alaska Fairbanks, USA

Contributing Authors
Elizabeth Bush, Environment Canada, Canada
Terry V. Callaghan, Abisko Scientific Research Station, Sweden; Sheffield Centre for Arctic Ecology, UK
Robert Corell, American Meteorological Society and Harvard University, USA
Shari Fox, University of Colorado at Boulder, USA
Christopher Furgal, Laval University, Canada

Alf Håkon Hoel, University of Tromsø, Norway
Henry Huntington, Huntington Consulting, USA
Erland Källén, Stockholm University, Sweden
Vladimir M. Kattsov, Voeikov Main Geophysical Observatory, Russia
David R. Klein, University of Alaska Fairbanks, USA
Harald Loeng, Institute of Marine Research, Norway
Marybeth Long Martello, Harvard University, USA
Michael MacCracken, Climate Institute, USA
Mark Nuttall, University of Aberdeen, Scotland, UK; University of Alberta, Canada
Terry D. Prowse, University of Victoria, Canada
Lars-Otto Reiersen, Arctic Monitoring and Assessment Programme, Norway
James D. Reist, Fisheries and Oceans Canada, Canada
Aapo Tanskanen, Finnish Meteorological Institute, Finland
John E. Walsh, University of Alaska Fairbanks, USA
Betsy Weatherhead, University of Colorado at Boulder, USA
Fred J. Wrona, National Hydrology Research Institute, Canada

Observers Accredited to the Arctic Council

Observer Countries:

France
Germany
The Netherlands
Poland
United Kingdom

International Organizations:

Conference of the Parliamentarians of the Arctic Region
International Federation of Red Cross & Red Crescent Societies (IFRC)
International Union for the Conservation of Nature (IUCN)
Nordic Council of Ministers (NCM)
Northern Forum
North Atlantic Marine Mammal Commission (NAMMCO)
United Nations Economic Commission for Europe (UN-ECE)
United Nations Environment Program (UNEP)
United Nations Development Programme (UNDP)

Non-Governmental Organizations:

Advisory Committee on Protection of the Seas (ACOPS)
Association of World Reindeer Herders
Circumpolar Conservation Union (CCU)
International Arctic Science Committee (IASC)
International Arctic Social Sciences Association (IASSA)
International Union for Circumpolar Health (IUCH)
International Work Group for Indigenous Affairs (IWGIA)
University of the Arctic (UArctic)
Worldwide Fund for Nature (WWF)

External Reviewers for *Impacts of a Warming Arctic*

Robert White, Washington Advisory Group, USA
Randy Udall, Community Office for Resource Efficiency, Aspen, Colorado, USA
Rasmus Hansson, World Wildlife Federation, Norway
Mary Simon, Former Ambassador for Circumpolar Affairs and Consultant, Canada
Ted Munn, University of Toronto, Canada
Roger G. Barry, National Snow and Ice Data Center, University of Colorado at Boulder, USA
O.W. Heal, University of Durham, UK

ASSESSMENT STEERING COMMITTEE

Representatives of Organizations

Robert Corell, Chair	International Arctic Science Committee, USA
Pål Prestrud, Vice-Chair	Conservation of Arctic Flora and Fauna, Norway
Snorri Baldursson (to Aug. 2000)	Conservation of Arctic Flora and Fauna, Iceland
Gordon McBean (from Aug. 2000)	Conservation of Arctic Flora and Fauna, Canada
Lars-Otto Reiersen	Arctic Monitoring and Assessment Programme, Norway
Hanne Petersen (to Sept. 2001)	Arctic Monitoring and Assessment Programme, Denmark
Yuri Tsaturov (from Sept. 2001)	Arctic Monitoring and Assessment Programme, Russia
Bert Bolin (to July 2000)	International Arctic Science Committee, Sweden
Rögnvaldur Hannesson (from July 2000)	International Arctic Science Committee, Norway
Terry Fenge	Permanent Participants, Canada
Jan-Idar Solbakken	Permanent Participants, Norway
Cindy Dickson (from July 2002)	Permanent Participants, Canada

ACIA Secretariat

Gunter Weller, Executive Director	ACIA Secretariat, USA
Patricia A. Anderson	ACIA Secretariat, USA

Lead Authors

Jim Berner	Alaska Native Tribal Health Consortium, USA
Terry V. Callaghan	Abisko Scientific Research Station, Sweden
	Sheffield Centre for Arctic Ecology, UK
Henry Huntington	Huntington Consulting, USA
Arne Instanes	Instanes Consulting Engineers, Norway
Glenn P. Juday	University of Alaska Fairbanks, USA
Erland Källén	Stockholm University, Sweden
Vladimir M. Kattsov	Voeikov Main Geophysical Observatory, Russia
David R. Klein	University of Alaska Fairbanks, USA
Harald Loeng	Institute of Marine Research, Norway
Gordon McBean	University of Western Ontario, Canada
James J. McCarthy	Harvard University, USA
Mark Nuttall	University of Aberdeen, Scotland, UK
	University of Alberta, Canada
James D. Reist (to June 2002)	Fisheries and Oceans Canada, Canada
Fred J. Wrona (from June 2002)	National Water Research Institute, Canada
Petteri Taalas (to March 2003)	Finnish Meteorological Institute, Finland
Aapo Tanskanen (from March 2003)	Finnish Meteorological Institute, Finland
Hjálmar Vilhjálmsson	Marine Research Institute, Iceland
John E. Walsh	University of Alaska Fairbanks, USA
Betsy Weatherhead	University of Colorado at Boulder, USA

Liaisons

Snorri Baldursson (Aug. 2000 - Sept. 2002)	Conservation of Arctic Flora and Fauna, Iceland
Magdalena Muir (Sept. 2002 – May 2004)	Conservation of Arctic Flora and Fauna, Iceland
Maria Victoria Gunnarsdottir (from May 2004)	Conservation of Arctic Flora and Fauna, Iceland
Snorri Baldursson (from Sept. 2002)	Arctic Council, Iceland
Odd Rogne	International Arctic Science Committee, Norway
Bert Bolin (to July 2000)	Intergovernmental Panel on Climate Change, Sweden
James J. McCarthy (June 2001 – April 2003)	Intergovernmental Panel on Climate Change, USA
John Stone (from April 2003)	Intergovernmental Panel on Climate Change, Canada
John Calder	National Oceanic and Atmospheric Administration, USA
Karl Erb	National Science Foundation, USA
Hanne Petersen (from Sept. 2001)	Denmark

Appendix - 3

Illustration and Photography

Project Production, Design and Lay-out:

Grabhorn Studio, Inc., 1316 Turquoise Trail, Cerrillos, New Mexico, 87010
United States (505) 780-2554 - grabhorn@earthlink.net

Graphics and Illustrations:

Inside Cover: arctic map - ©Clifford Grabhorn
All map backgrounds and map visualizations - ©Clifford Grabhorn/Grabhorn Studio, with the exception of those listed below.
Page 2: globe background - NASA
Page 25: sea ice extent images - NASA
Pages 32 - 33: flat globe map background - NASA
Page 54: spruce beetle map Yukon - Natural Resources Canada, spruce beetle map Kenai Peninsula - USDA Forest Service
Page 109: ©UNEP

The graphics are originals or based on preparatory files supplied by the individuals or institutes listed on the authorship page at the front of this book. References to original sources are given with the corresponding figures in the ACIA full Science Report.

Photography

Cover: all photographs - ©Bryan and Cherry Alexander, Higher Cottage, Manston, Sturminster Newton, Dorset DT10 1EZ, England - alexander@arcticphoto.co.uk
Title Page: ©Paul Grabhorn
Preface: ©Bryan and Cherry Alexander
Contents: all - ©Bryan and Cherry Alexander
Page 2: globe background - NASA
Page 4: earth images - NASA
Pages 6 - 9: all - ©Bryan and Cherry Alexander
Pages 10 - 11: ocean inundation at Shishmaref - ©Tony Weyionanna, all others - ©Bryan and Cherry Alexander
Page 12: river and snow landscape - ©Bryan and Cherry Alexander, permafrost - ©Paul Grabhorn
Page 13: river ice and sea ice with boat - ©Bryan and Cherry Alexander, glacier - ©Paul Grabhorn, coastal erosion - ©Stanilas Ogorodov, Moscow University
Page 14: forest fire - BLM Alaska Fire Service, all others - ©Bryan and Cherry Alexander
Page 15: stratospheric clouds - NASA, old-growth forest - ©Robert Ott, tundra - ©Bryan and Cherry Alexander
Pages 16 - 17: garden - ©Paul Grabhorn, all others - ©Bryan and Cherry Alexander
Page 20: snow landscape inset - ©Bryan and Cherry Alexander, sea ice inset - NASA
Page 21: Ellesmere Island glaciers from space - NASA
Page 22 -24: all - ©Bryan and Cherry Alexander
Page 25: sea ice - ©Bryan and Cherry Alexander
Pages 30 - 31: sea ice with pressure ridge and snow covered trees - ©Bryan and Cherry Alexander
Pages 33 - 35: all - ©Bryan and Cherry Alexander
Pages 37 - 38: all - ©Bryan and Cherry Alexander
Page 39: lake and mountain - ©Paul Grabhorn, forest growth, fire damage, lakes and ponds inset photos - ©Robert Ott, tundra ponds - ©Paul Grabhorn, phytoplankton - NASA
Pages 40 - 41: ice sheet aerials - ©Bryan and Cherry Alexander, 1958 McCall glacier - ©Austin Post, 2003 McCall glacier - ©Matt Nolan
Pages 42 - 43: Shishmaref coastal - ©Tony Weyionanna, low-lying islands - ©Paul Grabhorn, swamp sunset - US Army Corps of Engineers
Page 44-45: bird in flight - ©Frank Todd/B&C Alexander, all others - ©Bryan and Cherry Alexander
Pages 46 - 47: coast of Iceland - ©Snorri Baldursson, polar desert, semi-desert, tussock tundra - ©Terry V. Callaghan, all others - ©Bryan and Cherry Alexander
Pages 48-49: meltpond in Sweden - ©Terry V. Callaghan, all others - ©Bryan and Cherry Alexander
Page 50: fall forest - ©Robert Ott, lake aerial - ©Bryan and Cherry Alexander
Page 52: - Siberian forest - ©Bryan and Cherry Alexander
Page 53: - spruce trees and mountain - ©Robert Ott
Pages 54 - 55: spruce bark beetle - The National Agricultural Library Special Collections, spruce budworm - ©Therese Arcand/Natural Resources Canada, budworm infestation - ©Claude Monnier/Natural Resources Canada, spruce trees and mountainside - ©Robert Ott
Page 56: forest fire - ©John McColgan/BLM Alaska Fire Service
Page 57 - 59: all - ©Bryan and Cherry Alexander
Page 60: ice algae and diver - ©Rob Budd/NIWA
Pages 61 - 65: all - ©Bryan and Cherry Alexander
Pages 66 - 67: aquaculture in the Faroe Islands - ©Jens Kristian Vang
Page 69: caribou - ©Bryan and Cherry Alexander
Pages 70 - 71: cooking caribou - ©Henry Huntington, all others - ©Bryan and Cherry Alexander
Page 72: caribou migration aerial - ©Bryan and Cherry Alexander, Old Crow meeting and aerial - ©Paul Grabhorn
Page 73: caribou leaving river - ©Bryan and Cherry Alexander, five images of preparing caribou - ©Tookie Mercredi
Page 74: river aerial - ©Bryan and Cherry Alexander
Page 75: Tanana river - ©Robert Ott
Pages 76 - 77: all - ©Bryan and Cherry Alexander
Pages 78 - 79: St. George - , Nelson Lagoon -
Pages 80 - 81: Shishmaref storm and embankment - ©Tony Weyionanna, storm waves in Tuktoyaktu - ©Steve Solomon, coastal erosion and oil storage - ©Stanilas Ogorodov, Moscow University
Pages 82 - 83: all - ©Bryan and Cherry Alexander
Pages 84 - 85: oil spill images - Exxon Valdez Oil Spill Trustee Council, all others - ©Bryan and Cherry Alexander
Page 86: stuck truck - ©Paul Grabhorn, ice road - ©Bryan and Cherry Alexander
Page 88: ©Bryan and Cherry Alexander
Page 89: damaged building - ©Vladimir E. Romanovsky, BP building - ©Bryan and Cherry Alexander, avalanche -
Pages 90 - 91: all - ©Paul Grabhorn
Pages 92 - 93: drumming image - ©Henry Huntington, all others - ©Bryan and Cherry Alexander
Pages 94 - 97: all - ©Bryan and Cherry Alexander
Pages 98: stratospheric clouds - NASA
Page 100: ice landscape - ©Henry Huntington, plants - ©Paul Grabhorn
Page 101: stratospheric clouds - NASA
Page 102: both - ©Bryan and Cherry Alexander
Page 103: nesting bird - ©Bryan and Cherry Alexander, three autumn moth damage images images - ©Staffan Karlsson
Pages: 104 - 111 - all - ©Bryan and Cherry Alexander
Page 114: alpine pond and meadow - ©Paul Grabhorn, all others - ©Bryan and Cherry Alexander
Page 115: Saami herder and reindeer - ©Bryan and Cherry Alexander, harbor and island - ©Snorri Baldursson
Page 116: all - ©Bryan and Cherry Alexander
Page 117: damaged building - ©Vladimir E. Romanovsky, reindeer herder - ©Bryan and Cherry Alexander
Page 118: salmon fisherman and aerial landscape - ©Paul Grabhorn, Alaska landscape and oil tanker - ©Bryan and Cherry Alexander
Page 119: both - ©Bryan and Cherry Alexander
Page 120: top - NASA, migrating caribou and seal - ©Bryan and Cherry Alexander
Page 121: both - ©Bryan and Cherry Alexander
Pages 122 - 123: NASA
Pages 124 - 125: all - ©Bryan and Cherry Alexander
Back Cover: ©Bryan and Cherry Alexander